Robert U. Ayres

TECHNOLOGICAL
FORECASTING
AND
LONG-RANGE PLANNING

McGRAW-HILL BOOK COMPANY
New York   St. Louis   San Francisco   London   Sydney
Toronto   Mexico   Panama

02663

1234567890 MAMM 754321069

PREFACE

This book has grown out of an interest in technological forecasting which dates back nearly a decade. In that time I have exchanged ideas with a great many individuals, some of whom will inevitably fail to be adequately recognized. Among the people who must not be overlooked, however, are Raymond Gastil, Herman Kahn, Felix Kaufman, Max Singer, and Anthony Weiner, all former colleagues at Hudson Institute. I should also particularly mention Erich Jantsch of the OECD; Alan Kneese (and many others) of Resources For the Future Inc., Alan Berman, formerly Director of Hudson Laboratories (now Director of the Naval Research Laboratories); Col. Raymond Isensen, formerly in the Office of the Director of Defense Research and Engineering; and Theodore Taylor, President of International Research and Technology Corporation.

The book was partly written as a personal venture while I was on the Research Staff, and later a consultant, of the Hudson Institute; it was completed during a stint with Resources For the Future Inc. The original manuscript was prepared by Mrs. Lillian Drescher; the final version by Miss Dee Stell. Proofreading and indexing assistance have been provided by Mr. Richard McKenna and Mrs. Judith Wickham.

For the encouragement and many contributions of these people I would like to express my sincere thanks. For errors, inadequacies, or infelicitous adaptions of the good ideas of others, the author is solely responsible.

Robert U. Ayres

CONTENTS

vii

GLOSSARY

ANALOGY   recognizable similarity or resemblance of form or function, but no logical connection or equivalence—as distinguished from a *model*. (Synonymous with *metaphor* in the present context.)

ANALYTIC (MODEL)   in which operational outcomes are predicted from fundamental laws or first principles.

BENEFIT-COST (ANALYSIS)   a method of weighing the relative utility of various alternative courses of action, or investments, emphasizing a comparison of projected benefits and costs, reduced to common units (usually dollars).

CONJECTURE   surmise, guess, estimate, judgment from incomplete evidence.

CONTINGENCY TREE   a graphical display of logical relationships among actions or events focusing on branch points where several alternative outcomes are possible.

COST-EFFECTIVENESS (ANALYSIS)   an aspect of *systems analysis* concerned with quantifying the tradeoffs between the cost of a project and the quality of the results.

DECISION TREE   essentially a *contingency tree,* emphasizing branch points where decisions must be made.

DIFFUSION (OF A NEW TECHNOLOGY)   the evolutionary process of replacement of an old technology by a newer one for solving similar problems or accomplishing similar objectives.

DISCOVERY   a previously unknown datum or observed natural phenomenon (as contrasted with *invention*).

EMPIRICAL (MODEL)   where operational behavior is predicted from ad hoc relationships based on evidence from direct observation of natural phenomena or experimental data.

ENDOGENOUS (VARIABLE)   one whose dynamical behavior is encompassed within, understandable in terms of, and presumably derivable from, a specific *model, paradigm,* or discipline being considered.

EXOGENOUS (VARIABLE)   one whose behavior is not explainable in terms of the given model, etc.

EXPLORATORY (FORECAST)   one in which hypothetical future consequences of existing trends are exhibited from the standpoint of a neutral observer or nonparticipant.

EXTRAPOLATION (OF TRENDS)   estimation of future values or magnitudes assuming smooth, continuous progression.

FIGURE OF MERIT   an index of performance or utility.

FORECAST   a reasonably definite statement about the future, usually qualified in the sense of being contingent on an unchanging or very slowly changing environment (e.g., no wars or depressions).

GAME   a structured exercise involving any number of players each of whom seeks to maximize some objective function, subject to a set of rules. The objective functions and rules for different players are not necessarily the same and there may or may not be benefits from cooperation among players.

HEURISTIC (FORECAST)   as defined in the text—a forecast based on a *model* of the future, involving a greater degree of understanding than a simple extrapolation.

INNOVATION   the introduction or application of a new idea or invention; change in the existing order.

INVENTION   a new device, mechanism, or contrivance conceived by the human brain (as contrasted with a *discovery*).

METAPHOR   in the context of this book, synonymous with *analogy*.

MISSION TAXONOMY   a logical breakdown showing the logical interrelationships among missions from greater to lesser inclusiveness.

MODEL   an intellectual construct bearing some relation to reality, which can be discussed and analyzed in and of itself.

MORPHOLOGICAL METHOD   an analytical method emphasizing fundamental structural differences and/or similarities rather than functional or performance features.

NORMATIVE (FORECAST)   a forecast based on goals, purposes, objectives, or needs; often self-fulfilling by intention.

ONTOLOGICAL (VIEW)   the viewpoint in which processes of technological change are interpreted as self-generating or "inner-directed" (as contrasted with the *teleological* view).

PARADIGM   a structured set of axioms, assumptions, concepts, hypotheses, models, and theories, e.g., Newtonian physics or Marxist economics.

PENETRATION (OF TECHNOLOGY)   synonymous with *diffusion* as used in this book.

PERT   Program Evaluation and Review Technique.

PHENOMENOLOGICAL (MODEL)   a model which encompasses a set of observed phenomena into a formalism capable of being used for prediction, yet which does not purport to explain the underlying causes or relate them to general laws.

PLAN   an ordered, definite arrangement or sequence of actions to achieve a specified objective or target.

PPBS   Planning-Programming-Budgeting System first used in the U.S. Department of Defense, now practiced throughout the federal government.

PREDICTION   an unqualified apodictic statement about a future or as yet unobserved event (as contrasted with a *projection*).

PROGRAM   a menu of ideas, plans, and projects designed to further a goal.

PROJECTION   a contingent statement about the future, usually one of a group covering a range of possibilities.

QUASI MODEL   a conceptualization with more predictive power than an analogy or metaphor, but less quantification or precision than a model.

RELEVANCE TREE   a logical network similar to a *contingency tree,* but designed explicitly to elucidate the degree of importance of various "inputs" (e.g., projects) to a broadly defined outcome or goal.

SCENARIO   a time-ordered, episodic sequence of events bearing a logical (cause-effect) relationship to one another and designed to illumine a hypothetical future situation. If the moves of various players in a game (with rules designed to simulate real-world situations) are recorded, the results will be a scenario of sorts. Evidently the Hollywood version of the term is consistent with the more general usage. A scenario is not and is not intended to be either a *prediction* or a *forecast.*

TELEOLOGICAL (VIEWPOINT)   the view that processes of technological change should be interpreted as being responses to external stimuli (needs, demands, purposes, objectives), viz, "other directed" (as contrasted with the *ontological* view).

TRANSFER (OF TECHNOLOGY)   the application of a technology in a field *outside* the one for which it was developed or to which it was first applied (as contrasted with *diffusion* or *penetration*).

**TECHNOLOGICAL
FORECASTING
AND
LONG-RANGE PLANNING**

1

INTRODUCTION AND
HISTORICAL BACKGROUND

Knowledge of the future has been of obvious value since the dawn
of history. The discovery of subtle regularities in the apparent
motion of the stars and observed correlations between the positions
of the constellations and the annual floods in the valley of the
Tigris and Euphrates Rivers gave the possessors of this knowledge—
the priesthood—enormous economic power and wealth, and political influence. This spectacular success did not result in any great
advances in science generally, or forecasting in particular, although
it did lead to the establishment of empirical astronomy. Popular
astrology, the widespread superstition that human affairs are governed by the motions of the stars, is also apparently derived from
the preoccupations of the Mesopotamian priesthood.

Except insofar as they facilitate navigation, however, predictions
based on celestial mechanics have not been of great importance
since ancient times. Indeed for millennia, while the demand for
useful predictions remained high for obvious reasons, there was

no reliable source of knowledge about the future. To meet this difficulty, resourceful priests and oracles developed equivocation into a high art.[1] Nostradamus, the most famous seer of the Middle Ages, wrote verses so obscure and allusive that some people are still arguing over what he meant—though a better question might be whether he really intended to convey any information at all. However, although many oracles evidently made a career of deliberate ambiguity, there were others whose prophecies did have meaning and value. For the most part, nevertheless, these were intended to persuade and to encourage specific actions or policies. The examples which have been recorded and passed down to us tend to be those which were most successful as propaganda rather than those which reflected the best neutral judgment of situations and trends.

Skipping down to the present day, it is clear that whereas our ancestors generally lived in a relatively static society—with only the vaguest insight into what changes the future might hold—our own lives are embedded in a continuum with a nexus of concerns extending several decades into the future as well as the past. While most of us admit we know little or nothing about some aspects of the future—its political alignments or cultural fashions, for instance—we believe we know a lot about a number of other things. The "average" American, for instance, knows with high confidence:

1. Where he will be living and working and what he will be earning in five years
2. When the children will go to college and roughly how much it will cost him
3. When the mortgage on the house will be paid off
4. When he will retire and where he will go to live and how much he will have to live on
5. Approximately how long he will live (and by how many years his wife will outlive him)
6. What domestic political system he will live under, what the

[1] For instance, the oracle of Delphi was questioned by King Croesus of Lydia with regard to the outcome of his proposed attack on the Persian Empire. He was told in effect: "If you fight the Persians, a great empire will fall." Not surprisingly, the oracle proved to be correct.

major contending political philosophies will be, and how he will vote on most issues

7. The major international actors, what political system they represent (e.g., Communist or "free"), which ones will be strongest, richest, most threatening, and/or most friendly

As businessmen, bankers, actuaries, or government officials we can make quite good aggregate estimates of such things as the future labor force, employment level, demographic distribution, birthrate, annual inflation rate, gross national product, life expectancy, agricultural production, highway death rate, and demand for housing, fuel, electricity, transportation, and education. This knowledge is constantly put to practical use in both private and national affairs. Businessmen use demographic projections to forecast relative demand for various products (for instance, baby carriages versus wheel chairs) and to guide capital investment decisions and marketing strategies. Actuaries use life expectancies to calculate insurance premiums. Central bankers and economists use data on employment, gross national product, investment, trade, etc., to determine interest rates and monetary policy. Private decisions to buy stocks, bonds, gold and silver, or real estate are based on expectations about the future.

In short, forecasting, both explicit and implicit, is deeply woven into the fabric of twentieth-century Western civilization.

The forecasting of technological as opposed to economic or demographic change is not yet so universally practiced. In part, this lag is due to a belated recognition of the extent of the impact of technological change on society, and in part it is due to a rather widespread notion that technological change is inherently unpredictable. The popular mystique of science as an institution outside society, with an incomprehensible internal dynamism relying mainly on the workings of creative genius—resulting in perpetually unexpected "breakthroughs"—is largely responsible.

In recent decades, fortunately, both sociologists and economists have begun to think of science and technology less as a phenomenon in itself and more as an important, even vital, feature of the larger society with a particularly important influence on social and economic change. And, as we shall argue again later, although technology is generally created in response to societal demands or "needs"—as expressed in a market of some kind—many subsequent societal requirements arise from second-order effects of the technology itself.   Thus the societal demand for mobility and convenience in transportation in some sense "created" the automobile. But the automobile, in large numbers, has created a galaxy of external effects from traffic congestion to smog to automobile graveyards.   These, in turn, have created a potential outlet for new technological solutions such as the electric car or the automatic passenger conveyor system.   But here, the point is simply that social and economic forecasting in general require more and more explicit technological inputs and have led, by degrees, to a need for methodologies for forecasting technological change per se.

At the same time, the growth of research and technology as an economic activity involving large numbers of people and major resources also means an ever more acute need for forecasts as an adjunct to planning.   While this need is perhaps subsidiary to the broader social one, it is reflected more immediately in a specific marketplace—namely the military.   Thus the administration of military research and development has played a disproportionate role in the emergence of technological forecasting as a serious professional activity.

The earliest attempts at technological forecasting have largely been individual speculations—with an emphasis on social consequences—carried out as a quasi-intellectual exercise, but often with a strong bias toward providing maximum entertainment value.   A whole literary genre, science fiction, has grown up to explore the human implications of science and of technology.   Most of this material is worthless as prophecy, yet one must recognize the seminal influence of Jules Verne, H. G. Wells, Karel Čapek, Aldous Huxley, Robert Heinlein, Hal Clement, and others.   Some themes, such as space travel, automation (robotics), and "brainwashing," have been so thoroughly thrashed out that one sometimes has a sense of *déjà vu* when parallel developments occur in the real world.

A great many interesting, and several useful, forecasts have been made to arouse interest in, and further the cause of, specific developments such as aircraft.   Many scientists or military men have gone so far as to describe detailed systems and missions—sometimes in technical reports, sometimes in the form of fiction. One famous example of this kind was a series of writings produced during and after World War I by Giulio Douhet, an Italian general, who prophesied that future wars would be won by air power (and described some of the tactics and targets of such wars) [1]. Douhet was wrong on many points, including his assessment of the effectiveness of strategic bombing, but he was one of the first to see clearly—and help to influence—the direction of future trends in warfare.

Another area which abounds in "special pleading" for the future is space travel and exploration. All the important features of interplanetary travel via rocket ships were described more than 40 years before the fact by Herman Oberth's two books on space travel published in the 1920s [2,3]. Arthur C. Clarke proposed the communications satellite in 1945, many years in advance of the actual event [4]. Again there is little doubt that the writings of Clarke and others were quite influential in creating and maintaining intense public interest in space exploration, without which it is doubtful whether a moon probe would have been undertaken so soon (if ever). Two of the more recent examples of propaganda prophecy in this field are *The Next 50 Years in Space* by Dandridge Cole [5] and *The Case for Going to the Moon* by Neil Ruzic [6].

There have also been numerous denunciatory (self-denying) forecasts where it is alleged that something is undesirable, impractical, or impossible. Thus George Orwell's *1984* [7] and Aldous Huxley's *Brave New World* [8] have helped to alert people against some of the dangers of a too potent technology, particularly in such areas as electronic eavesdropping, subliminal communication, and genetic engineering. Nevil Shute's *On the Beach* [9], Pat Frank's *Mr. Adam* [10], and others have painted graphic, if often inaccurate, pictures of the aftermath of a nuclear war. Specific scientific programs and/or proposals are frequently debunked as enthusiastically as they are boosted. Arthur C. Clarke has collected a number of examples of particularly bad forecasts, some of which are worth quoting [11].

The first was an essay by the astronomer Simon Newcomb, who stated that "the demonstration that no possible combination of known substances, known forms of machinery, and known forms of force can be united in a practical machine by which man shall fly long distances through the air, seems to the writer as complete as it is possible for the demonstration of any physical fact to be." Newcomb was convinced (as were Euler, Stokes, Kirchhoff, and Rayleigh before him) that the physics of lift and drag on finite three-dimensional bodies moving in a viscous fluid (air) were such as to rule out any possibility of powered flight by heavier-than-air craft. Although the relevant physical laws were previously known in general, the correct deduction (i.e., calculation) from them was not made until after a successful powered flight was actually demonstrated by the Wright brothers in 1903.

A second instance was occasioned by the publication of a book called *Rockets through Space* by Cleator in 1936. A review of the book, written by R. v. d. R. Wooley and published in *Nature,* dismissed the notion of space flight as "essentially impractical." In 1956, a year before the first Sputnik, Dr. Wooley was appointed Astronomer Royal. When he was interviewed by the press he confirmed his earlier opinion with the remark: "Space travel is utter bilge [12]." The author of this opinion became, ex officio, a member of the committee advising the government of the United Kingdom on space research.

Aeronautical engineer Nevil Shute Norway (later famous as author Nevil Shute) was chief calculator for the R-100 Airship and a cofounder of Air Speed Ltd., subsequently merged with DeHavilland's in 1940. In 1929 he was very much an optimist about the future of civil aviation, yet he firmly predicted that by 1980 commercial aircraft would be limited to a cruising speed of 110 to 130 mph, a range of 600 miles, and a payload capacity of 4 tons out of 20 tons total weight. (On the other hand, in a novel written in 1948 [13], he accurately anticipated the problems of catastrophic structural failure due to metal "fatigue" which later plagued the Comet and the Lockheed Electra, among other aircraft.)

There have been numerous well-known debunkers of atomic energy, notably the famous Lord Rutherford, but one quote is enough. J. B. S. Haldane wrote a curious little book in 1925 called *Callinicus: A Defense of Chemical Warfare,* which contains one of the unluckiest technological forecasts ever made [14]:

> If we could utilize the forces which we now know to exist inside the atom we would have such capacities for destruction that I do not know of any agency other than divine intervention which would save humanity from complete and peremptory annihilation. But . . . we cannot make apparatus small enough to disintegrate or fuse atomic nuclei any more than we can make it large enough to reach the moon. . . . To do this we should require to construct apparatus on the same infinitesimal scale as the structure of the chemical atom. . . . And the prospect of constructing such an apparatus seems to me so remote that, when some successor of mine is lecturing to a party spending a holiday on the moon, it will still be an unsolved (though not, I think, an insoluble) problem.

A final example of overhasty debunking was committed independently by the chief scientific advisors of the British and United States governments. Professor F. A. Lindemann (Lord Cherwell) told the House of Lords in summer 1945, and Vannevar Bush testified to the U.S. Senate a few months later, that intercontinental ballistic missiles could not be competitive with bombers in the foreseeable future. Bush made his statement very strong:

> In my opinion such a thing is impossible for many years. . . . I think we can leave that out of our thinking. I wish the American public would leave that out of their thinking [15].

The reasons given were that most of the weight of a long-range rocket would perforce be needed for fuel; destructive capacity even of atomic warheads would be tightly constrained by limited payloads; and accuracy could not be increased to the point where there would be an acceptable probability of destroying a target several thousand miles away. Both men were an order of magnitude too pessimistic about accuracy, but the real "mistake" was not foreseeing even the possibility of the H-bomb and the remarkable progress in warhead miniaturization (kilotons per pound). On cost-effectiveness grounds the Lindemann-Bush 1945 projection was the logical one, considering the information available at the time. But by 1951, following Edward Teller's so-called "breakthrough," the H-bomb development was on the immediate horizon and a revision of the earlier conclusion would have been appropriate. However, this did not happen until 4 or 5 years later.

One of the earliest "neutral" attempts to forecast future technology in general terms was H. G. Wells's *Anticipations,* written in 1902 [16]. Wells, like other forecasters, made both good and bad predictions, some of which are interesting to compare with present-day reality. His worst forecasts concerned the airplane and the submarine. As to the former: "Aeronautics will never come into play as a serious modification of transport and communications [page 208]," although: "Very probably before 1950 a successful aeroplane will have soared and come home safe and sound [page 208]." As regards the latter, his "imagination refuses to see any sort of submarine doing anything but suffocating its crew and foundering at sea [page 217]," and "you may throw out a torpedo or so, with as much chance of hitting anything vitally as you would have if you were blindfolded [page 218]." His prognostications with respect to land travel were somewhat better: the advent of motor trucks and buses for heavy traffic transport and privately owned motor carriages with private (i.e., segregated) roads for high-speed motor travel were foreseen. A particularly significant remark is the following: "Only a revolutionary reconstruction of the railways or the development of some competing method of land travel can carry us beyond 50 miles an hour on land." Presumably he was speaking of average speeds. Except for a very small number of trains and a few high-speed turnpikes, where average speeds of 60 to 80 mph may be maintained over moderate distances, Wells was astonishingly accurate in this regard.[2] In 1911 Thomas Alva Edison, the father of so many electrical inventions, ventured several broad forecasts published in popular magazines from 1910 to 1914, mostly concerning future applications of electricity [17]. Among other things, Edison was a strong (and active) proponent of the electric car which, after a long season of drought, begins to look like it may make a comeback in the last quarter of this century. In 1911 Charles Steinmetz

[2] Although possibly for the wrong reasons.

of General Electric wrote an article for the *Ladies Home Journal* discussing the application of electricity to housekeeping [18]. Among his successful predictions were electric kitchens, air conditioning, mine-head power stations, legislation controlling the use of "fire" (i.e., combustion) in cities, and radio broadcasting.[3]

In October, 1920, *Scientific American* published an editorial article by Austin C. Lescarboura which made some 65 definite predictions aiming at the year 1995 or earlier [19]. There were major omissions, such as radio broadcasting (as well as radio telephony) and talking pictures, but only 15 years later, 25 predictions (38 percent) had already come true and 13 more (20 percent) looked very likely, while only 7 looked very unlikely to S. C. Gilfillan [20]. This characteristic telescoping of time frequently makes the earlier predictions appear to be too conservative.

In 1924 J. B. S. Haldane wrote *Daedalus, or Science and the Future* [21]. A few random selections are interesting: "In 50 years light will cost about a fiftieth of its present price and there will be no more night in our cities." And "We are working towards a condition when any two persons on earth will be able to be completely present to one another in not more than $\frac{1}{24}$ of a second. We shall never reach it, but we shall approach it indefinitely."

As regards power, Haldane looked to the eventual utilization of wind and sunlight, relying on electrolysis of water for energy storage and explosive recombination as a means of recapturing the energy. He felt liquid hydrogen would be used to power airplanes because of its high energy content per unit weight (it is being used for rockets for this reason). Haldane suggested that "Before long someone may discover that frescoes inside a factory increase the average efficiency of the worker 1.03% and art will become a commercial proposition once more" (as it has). However, his main interest lay in the field of biological engineering, for which he predicted a spectacular future well within the twentieth century, including total conquest of disease, systematic application of eugenics, the "invention" of new biological species—e.g., of algae—resulting in a catastrophic food *surplus,* a declining natu-

---

[3] Oddly enough, though many foresaw radio telephony, television, and other applications, broadcasting was not widely anticipated.

ral birthrate compensated by widespread artificially induced ecto-
genesis,[4] and so on.   On the whole, Haldane was a much worse
prophet than Wells.

Bertrand Russell in the same series of mythologically titled publications wrote (also in 1924) *Icarus, or the Future of Science* [22].
Russell foresaw an increase of political and economic centralization
and control, brought about by improved means of transportation
and communication (something like George Orwell's vision).   He
also thought that birth control would lead to stationary or declining
populations among the white nations by 1975:

> This situation will lead to a tendency—already shown by the
> French—to employ more prolific races as mercenaries.   Govern-
> ments will oppose the teaching of birth-control among the
> Africans for fear of losing recruits.   The result will be an im-
> mense numerical inferiority of the white races leading probably
> to their extermination in a mutiny of mercenaries. . . .

Russell also looked for governments to adopt eugenic principles
to eliminate not only "undesirables" but also internal opposition,
for widespread use of applied psychology to induce irrational beliefs
(i.e., that "governments are wise and good"), and, in general,
to use technology to increase their power over the average citizen.
Russell seems to have anticipated—if not stimulated—the gloomy
literary visions of Orwell and Huxley, but in other respects his
foresight was badly off target.

Alfred M. Low's book, *The Future* (1925) [23], sticks closer
to technology, although he too predicted that most phases of life
in the future would be controlled by the government and that
children would be educated and cared for by the state.   Low was
as fascinated by radio as Haldane was fascinated by biology, and
predicted artificial light by radio oscillation, "radio tape machines"
to record news received while the listener is absent, home "radio
therapy" for many diseases, and electric vehicles powered by rf
power supplied to cables buried in the "dustless and resilient"
roads.   He wrongly thought that battleships would evolve into
massive "floating forts," yet he also foresaw aircraft being launched
by compressed air (this should have suggested the floating airfield

---

[4] Conception and growth of the fetus outside the womb.

or aircraft carrier, which shortly superseded the battleship). One of his better insights, on the other hand, was that the future of chemistry would be along the lines of "atomic disintegration."

In 1936, C. C. Furnas, who later became an Assistant Secretary of Defense for research and development, wrote *The Next Hundred Years* [24]. In many ways this was the least unsatisfactory technological forecast (except for Steinmetz) of the pre-World War II era. Furnas foresaw enormously effective medicines to counteract infectious diseases such as pneumonia (i.e., antibiotics) and looked for similar progress in the war against insect pests, first via chemical insecticides and ultimately by biological controls. The social importance of television was recognized, though Furnas was skeptical about the social demand for such a device—his most noteworthy failure of insight. Transatlantic telephoning via undersea cables was foreseen (if satellites were not), as was a big future for pipeline distribution systems. As regards atomic energy, "it is quite evident that if we are going to get at the atomic energy idea, something radically different must be devised. Some way must be contrived *to make the atoms smash themselves, and release their energy spontaneously.*" This remark no doubt had its origin in Joliot-Curie's Nobel prize lecture of 1935, but boldly contradicted the great authority of Lord Rutherford. It anticipated the landmark Hahn-Strassman experiment by 3 years.

Many interesting individual efforts have been made since World War II. The outstanding examples are probably Sir George Thomson's *The Foreseeable Future* (1955) [25] and Arthur C. Clarke's *Profiles of the Future* (1962) [26]. The value of these and many other recent attempts to look ahead cannot be adequately assessed at this point, since too little time has elasped since their publication.

The "modern era" of systematic technological forecasting can be said to have begun in the mid-thirties when the National Resources Committee was set up by the National Research Council under the chairmanship of William F. Ogburn, who was professor of sociology at the University of Chicago. Its report, entitled *Technological Trends and National Policy,* was published in 1937 [27]. The justification for this effort was succinctly stated by Ogburn in the opening paragraphs of the Committee's report:

In an age of great change, anticipation of what will probably happen is a necessity for the executives at the helm of the ship of state. A study of invention offers a very good clue to future social conditions and problems of a nation. For, of four material factors that determine the economic well-being of nations, to wit, invention, population, natural resources, and economic organization, the first changes the most frequently in the modern world and hence is most often a cause.

In short, the National Research Council recognized that intelligent long-range planning requires insight into the social, technological, and military environments which will exist in the future.[5] Yet this study failed to foresee atomic energy, radar, antibiotics, or jet propulsion, all of which were under high-priority engineering development or in practical use 5 years later. The beginnings of radar go back to the late twenties in both the United States and the United Kingdom. The so-called Tizard Committee in England recommended a crash program of radar development for air defense, and this was already well under way at the time of publication of the Ogburn report. Penicillin had been discovered in 1929 and only required development of mass-production techniques. Jet engines were also well known in theory and awaited development of turbines and compressors capable of operating at high temperatures. In the case of atomic energy, perhaps, the future was not quite so readily foreseeable despite the optimism of Joliot-Curie, Szilard, and others.

Since World War II there have been a number of major collective efforts at forecasting, both in and out of government. Major efforts worth mentioning are the so-called Von Karman report (1944) [28] by the Scientific Advisory Board of the Air Force (which was one of the progenitors of the RAND Corporation). In the early fifties the Air Force Office of Scientific Management utilized technological forecasts as a guide to formulating research and development objectives. This led to a larger effort (1963–1964) called Project Forecast [29].

The U.S. Army has also been utilizing technological forecasts more or less continuously since 1957. The Army Materiel Com-

---

[5] An essay in the report, "The Prediction of Inventions," by S. C. Gilfillan, reviews a tremendous body of earlier forecasting literature quite thoroughly.

mand now issues a long-range forecast each year [30]. The Naval
Materiel Command also began a technological forecasting effort
in late 1965, which was formalized into an ongoing program by
early 1967 [31].

Other major efforts in the military area include a forecast of
science and technology to 1985 done by the University of Syracuse
Research Corporation for the U.S. Marine Corps (1964) [32],
and a survey of future military technology by Hudson Institute
jointly for the Arms Control and Disarmament Agency (ACDA)
and the Advanced Research Projects Agency (ARPA) in the Defense Department (1965) [33]. Most of the foregoing are, unfortunately, for restricted distribution only.

Nonmilitary forecasts are less numerous and, on the whole, less
systematic. A few examples worth mentioning are *Resources for
Freedom,* The Report of the President's Materials Policy Commission (known as the Paley Commission) in 1952, which contained
detailed forecasts of requirements and availability of key raw materials, energy sources, and materials technology [34]. The *National
Power Survey,* published (1964) by the FPC, contains a detailed
forecast of the electrical power industry to 1980 [35]. A somewhat
similar survey of future availability of petroleum fuels is being
carried out by the Bureau of Mines and the Office of Oil and
Gas in the Department of the Interior. The Bureau of Labor
Statistics in the Department of Labor has recently completed a
study of future technological trends in major United States industries and their impact on wages, labor requirements, and availability [36]. The bureau now has a continuing program of forecasting
the impact of technological change on the labor force.

Other agencies actively concerned with the impact of newly developing technologies on the future environment include the President's Science Advisory Committee, the Federal Council for Science
and Technology, and the Office of Science and Technology, the
National Science Foundation, the Institute of Applied Technology
of the National Bureau of Standards, the Bureau of Mines, the
AEC, and NASA, as well as temporary groups commissioned for
practical purposes [37].

Outside the federal government, programs worth mentioning
include the Twentieth Century Fund study *America's Needs and*

14   **Technological Forecasting**

Resources, by J. Frederick Dewhurst et al. [38] and a study initiated and carried out by members of the California Institute of Technology staff (1957) [39]. The work of the Paley Commission mentioned previously resulted in the establishment of an ongoing program under the aegis of a nonprofit corporation established for the purpose and funded by the Ford Foundation. Resources for the Future, Inc. has carried out nearly 100 major studies in these fields [40]. The Aerospace Industries Association prepared a 10-year forecast for the industry (1962) [41]. The American Academy of Arts and Sciences has recently completed a 2-year study of the year 2000 [42,43]. Abroad, Bertrand de Jouvenal's "Futuribles" program sponsored by Ford Foundation [44] and Robert Jungk's "Mankind 2000" project [45] also deserve mention. International organizations such as OECD, Euratom, CECA (Coal and Steel Community), ICAO (International Civil Aviation Organization), and others have also been prolific forecasters.

As regards private industry, more than half of the 500 largest corporations in the United States now have formal long-range planning programs. Most of these involve at least some degree of technological forecasting, although such activities tend to be limited as a rule to short-range goals (3 to 5 years) with informal forecasts up to 10 years in some cases. An exception is the "MIRAGE 75" study of Lockheed Aircraft Corporation (1964) [46]. Industry is beginning to take longer-range technological forecasting quite seriously; a successful conference was held in May 1967 under the auspices of the Industrial Management Center at Lake Placid, New York [47]. Two more conferences have been scheduled for 1968. This appears likely to become an annual affair. A more complete survey of forecasting efforts both in government and industry has been prepared by Erich Jantsch, a consultant to the OECD. This book constitutes the most complete available source of bibliographical and "state-of-the-art" material [48].

REFERENCES

1. Giulio Douhet, *The Command of the Air,* trans. by Dino Ferrari, Coward-McCann, Inc., New York, 1942.

2. H. Oberth, *Die Rakete zu den Planetenraumen,* R. Oldenbourg KG, Munich, 1923.

3. H. Oberth, *Wege zur Raumschiffahrt,* R. Oldenbourg KG, Munich, 1929.

4. Arthur C. Clarke, "Extraterrestrial Relays," *Wireless World,* London, October, 1945.

5. Dandridge Cole, *Beyond Tomorrow: The Next 50 Years in Space,* Amherst Press, Amherst, Wis., 1965.

6. Neil Ruzic, *The Case for Going to the Moon,* G. P. Putnam's Sons, New York, 1965.

7. George Orwell (Eric Blair), *1984,* Harcourt, Brace & World, Inc., New York, 1949.

8. Aldous Huxley, *Brave New World,* Doubleday & Company, Inc., Garden City, N.Y., 1932.

9. Nevil Shute (N. S. Norway), *On the Beach,* William Heinemann, Ltd., London, 1957.

10. Pat Frank, *Mr. Adam,* J. B. Lippincott Co., Philadelphia, 1946.

11. Arthur C. Clarke, *Profiles of the Future: An Enquiry into the Limits of the Possible,* Victor Gollancz, Ltd., London, 1962.

12. *Ibid.*

13. Nevil Shute (N. S. Norway), *No Highway,* William Heinemann, Ltd., London, 1948.

14. J. B. S. Haldane, *Callinicus: A Defense of Chemical Warfare,* Kegan Paul, London, 1925.

15. Vannevar Bush, testimony before the Senate Special Committee on Atomic Energy, December, 1945.

16. H. G. Wells, *Anticipations,* Harper & Brothers, New York & London, 1902.

17. Allan L. Benson, interview with Thomas Edison, "The Wonderful New World Ahead of Us," *Cosmopolitan Magazine,* vol. 50, p. 294, 1911.

18. Charles P. Steinmetz, "You Will Think This a Dream," *Ladies Home Journal,* Sept. 15, 1915.

19. A. C. Lescarboura, "The Future as Suggested by the Developments of the Last 75 Years," *Scientific American,* vol. 123, pp. 320–321, Oct. 2, 1920.

20. S. C. Gilfillan, "The Prediction of Inventions," from *National Trends and National Policy,* U.S. Natural Resources Council, 1937.

21. J. B. S. Haldane, *Daedalus, or Science and the Future,* Kegan Paul, London, 1924.

22. Bertrand Russell, *Icarus, or the Future of Science,* Kegan Paul, London, 1924.

23. Alfred M. Low, *The Future,* International Publishers Company, New York, 1925.

24. C. C. Furnas, *The Next Hundred Years: The Unfinished Business of Science,* The Williams & Wilkins Company, Baltimore, 1936.

25. Sir George Thomson, *The Foreseeable Future,* Cambridge University Press, New York, 1955.

26. Clarke, *Profiles of the Future, op. cit.*

27. W. F. Ogburn et al., *Technological Trends and National Policy,* U.S. National Research Council, Natural Resources Committee, 1937.
28. Theo von Karman et al., "Towards New Horizons," USAF Scientific Advisory Group, Nov. 7, 1944.
29. *Operation FORECAST* (SECRET), 14 vol., SCGF-46-7, Forecast Special Project Office, HQ AF Systems Command, January, 1964.
30. *Army Long-Range Technological Forecast* (SECRET, NOFORN), 4th ed., 3 vols. Army Materiel Command, 1967.
31. M. J. Cetron, R. J. Happel, W. C. Hodgson, W. A. McKenney, and T. I. Monahan, "A Proposal for a Navy Technological Forecast," 2 vols., HQ Naval Materiel Command, Washington, D.C., May 1, 1966.
32. "Science and Technology in the 1985 Era," supplement to "The United States and the World in the 1985 Era," prepared by the Physical and Social Science Faculties of the University of Syracuse, under the auspices of Syracuse University Research Corp., DDC Accession number AD 613526, Mar. 15, 1964.
33. D. G. Brennan (ed.), "Future Technology and Arms Control" (SECRET), Hudson Institute HI-504-RR, June, 1965.
34. *Resources for Freedom,* Report of the President's Materials Policy Commission, Washington, D.C., 1952.
35. *National Power Survey,* 2 vols., U.S. Federal Power Commission, 1964.
36. "Technological Trends in Major American Industries," *Department of Labor Bulletin* 1474, U.S. Bureau of Labor Statistics, 1966.
37. See *Technology and the American Economy* (summary volume plus 6 appendices), Report of the National Commission on Technology, Automation and Economic Progress, Washington, D.C., 1966. Also *Applied Science and Technological Progress,* a report to the Committee on Science and Astronautics, U.S. House of Representatives, by the National Academy of Sciences, 1967.
38. J. Frederick Dewhurst et al., *America's Needs and Resources,* Twentieth Century Fund, New York, 1947, 1955.
39. Harrison Brown, James Bonner, and John Weir, *The Next Hundred Years,* The Viking Press, Inc., New York, 1957.
40. Notably H. H. Landsberg, L. L. Fischman, and Joseph L. Fisher, *Resources in America's Future,* Resources for the Future, Inc., The Johns Hopkins Press, Baltimore, 1962. Resources for the Future has published a large number of other monographs many of which involve forecasts.
41. *Aerospace Technical Forecast 1962–1972,* Aerospace Industries Association (AIA), Washington, D.C., 1962.
42. "The Year 2000," *Daedalus* (special issue), Summer, 1967. Condensed from "Working Papers of the Commission on the Year 2000," 5 vols., American Academy of Arts and Sciences, Boston (not published).
43. Herman Kahn, and A. J. Wiener, *The Year 2000: A Framework for Speculation,* The Macmillan Company, New York, 1967. (Based on vol. II of the "Working Papers of the Commission on the Year 2000," ref. 42).
44. See, for instance, Bertrand de Jouvenal, *L'Art de la Conjecture,* Edition de Rocher, Monaco, 1964. English translation, Basic Books, Inc., Publishers, New York, 1967.

45. Institut für Zukunftsfragen, Vienna, Austria.
46. H. A. Linstone, "MIRAGE 75—Military Requirements Analysis Generation 1970–75," Report No. LAC/592371, Lockheed Aircraft Corp., Burbank, Calif. (SECRET RD). A revised version, "MIRAGE 80," is now in preparation.
47. See James Bright (ed.), *Proceedings of the 1st Annual Technology and Management Conference,* Prentice-Hall, Inc., Englewood Cliffs, N.J., 1968.
48. Erich Jantsch, *Technological Forecasting in Perspective,* OECD, Paris, 1967.

2 FAILURES OF TECHNOLOGICAL FORECASTING

Forecasting in general is beset by hazards for the would-be prophet. Most of these hazards—the uncertainty and unreliability of data, the complexity of "real world" feedback interactions, the temptations of wishful or emotional thinking, the fatal attraction of ideology or an *idée fixe*,[1] the dangers of forcing soft and somewhat pliable "facts" into a preconceived pattern—apply to all forms of forecasting. In addition, it appears, there are some pitfalls due to the special characteristics of invention and innovation as social processes (and perhaps of the people who do most of the crystal-ball gazing in this field). Some of these pitfalls are worth explicit recognition and a brief commentary.

1. *Lack of imagination and/or "nerve."* This is a particular problem of the committee of eminent experts, many of whom in-

[1] For instance, Marxist forecasting is often ludicrously off base because of its doctrinaire insistence on a very narrow and limited view of human nature and historical causality.

stinctively prefer to look too cautious than not cautious enough (especially to each other) even though they may be conscious of this pitfall and try honestly to compensate for it. An example may serve to emphasize the point. In 1940 the National Academy of Sciences appointed a select committee to evaluate the proposed gas turbine [1]. The membership of the committee included Theodore von Karman, Charles Kettering, Robert Millikan, Max Mason, A. G. Christie, and Lionel S. Marks. Their reasoned and careful conclusion, based on many conservative assumptions, was that gas turbines would have to weigh 13 to 15 lb/hp compared with 1.1 lb/hp for internal-combustion engines then operating. If the committee had been optimistic instead of pessimistic in its choice of assumption, it could have arrived at a figure of 0.4 lb/hp (which was the correct one). In point of fact, a gas turbine was in operation in England only a year later.

2. *Overcompensation.* There are many well-known cases of prophets and inventors being ignored in their own time and country, and later being stunningly vindicated, often by others: viz. the U.S. Air Force's Billy Mitchell; Charles de Gaulle, pioneer advocate of "blitzkrieg"; Frank Whittle and the turbojet engine; Tsiolkovsky, Oberth, and Goddard and rockets, etc. As a result, some people tend to lean over backward nowadays and say, in effect, that "no matter how fantastic our expectations may be, the truth will be more incredible." To quote Arthur C. Clarke:

> Anything that is theoretically possible will be achieved in practice, no matter what the technical difficulties, if it is desired greatly enough. It is no argument against any project to say: "the idea is fantastic!" Most of the things that have happened in the last fifty years have been fantastic, and it is only by assuming that they will continue to be so that we have any hope of anticipating the future [2].

Clarke, taking his own advice seriously, suggests in a "time-table for the future" that by 2050 we will have achieved gravity control and, by 2100, human immortality. Another (Soviet) optimist has written [3]: "There are no limits to creative human thinking. . . . In our day the human genius can do anything."

Several science popularizers of the "Gee Whiz" school have used the technique of "envelope curve" extrapolation in order to justify very radical predictions. One author remarks that the rates of increase of a number of performance variables apparently will go asymptotically to infinity before the year 2000 [4]. Thus:

   a. Using the envelope curve for vehicular speeds (Fig. 2.1) it appears that the speed of light would seem to be achieved by 1982. It is interesting to compare Fig. 2.1 with Fig. 2.2, which represents the same basic information plotted on a different scale, and leads to a very different sort of prediction [5].

   b. From life-expectancy trends he concludes that "anyone born after 2000 A.D. lives forever, barring accidents." (If this extrapolation is correct, Clarke is indeed too conservative. However, there is little or no indication that the *maximum* age for humans is increasing. In fact, it appears to remain quite constant at about 115, although more people nowadays have a chance of attaining this lifespan.)

   c. Using a trend curve similar to Fig. 2.3 [6], he argues that by 1981 "a single man will have available under his control the amount of energy equivalent to that generated by the entire sun [7]."

   d. Using another trend curve (not shown), he suggests that by 1970 the number of "circuits" in a computer may be equal to the number of neurons in the human brain, i.e., about 4 billion. The practical significance of this comparison is not clear, of course, though the notion is undeniably provocative.

   3. *Failure to anticipate converging developments* and/or changes in competitive systems. To cite a well-known example, nuclear power has advanced more slowly than was originally thought in the 1940s and at greater cost, primarily because of simultaneous (and largely unexpected) improvement in the efficiency of fossil-fuel power-generating systems, as shown in Fig. 2.4 [8]. Similarly, titanium and beryllium technology have lagged far behind the rosy expectations of a decade ago largely because the tremendous market for high strength-to-weight structural members for bombers such as the B-70—which would have justified heavy development costs in metallurgy and fabrication techniques—suddenly disappeared. The same fate overtook the re-

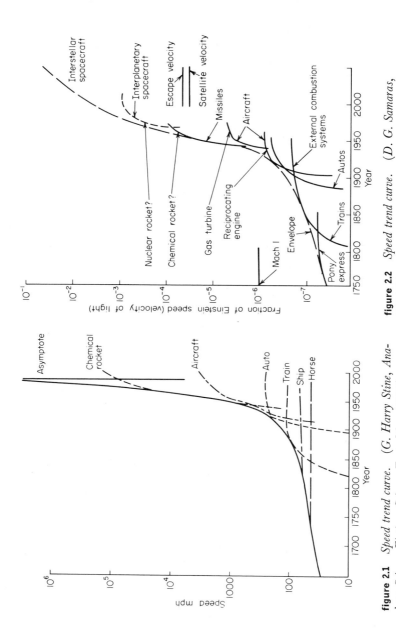

figure 2.1 *Speed trend curve.* (G. Harry Stine, Analog: Science Fiction, Science Fact, May, 1961.)

figure 2.2 *Speed trend curve.* (D. G. Samaras, USAF.)

The same parameter is plotted in both graphs. The one on the left is extremely misleading because an inappropriate scale is used.

21

search and development programs in high-energy (e.g., borate) fuels and nuclear-powered aircraft such as SLAM.

The problem of "blinders" is not unique to advanced forecasting technology.   Figure 2.5 summarizes several past forecasts of mass transit utilization in Chicago [9].   In each case the forecast was a simple extrapolation of a rising trend, which totally ignored rising competition from the automobile, shortened working hours, and the impact of the 5-day week.   An interesting example of a forecast which did *not* fail to consider converging developments is due to S. C. Gilfillan, who in 1913 forecast that the size of future ocean liners would not continue to increase as a simple extrapolation—shown in Fig. 2.6—but instead would reach a peak by 1925 followed by a sharp decline and a slower subsequent rise [10].   Gilfillan reasoned correctly that competition from aircraft would ultimately cut into the liners' business.

Persistent overestimation of the rate at which technological innovations can be put into practice (because of inertia, caution, long lead times, and a desire not to jeopardize existing investments) is as endemic among forecasters as persistent underestimation of the rate of future scientific progress.   As a result, progress in

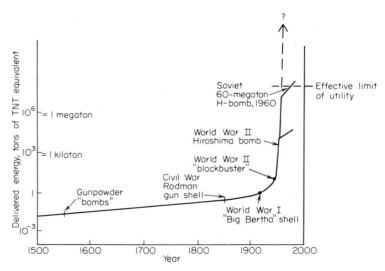

figure 2.3   *Explosive power trend curve.*

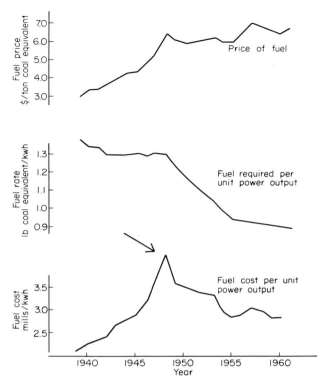

figure 2.4   *Fuel trends for conventional steam power plants. Note the unexpected reversal of the rising trend of fuel costs. This was the factor that kept conventional power from being replaced by nuclear power in the late fifties.*

science often exceeds our expectations while technology typically lags far behind them. Most people will recall a time shortly after World War II when it was confidently and widely expected that helicopters would shortly replace the family car. Other flashes in the pan of recent memory include radar ovens for virtually instantaneous cooking,[2] thermoelectric refrigerators for household use, plastic and/or fiberglass automobile bodies, light noncorroding metals such as magnesium, beryllium, and titanium to replace structural steel, and so forth.

---

[2] Now finding commercial applications in aircraft, however.

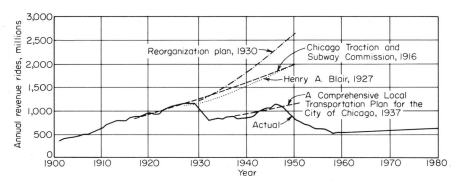

figure 2.5   *Actual use of transit facilities from 1901 to 1960 and esti-mated to 1980 compared with various projections of estimated use (Chicago).*

4. *Concentration on specific configurations,* rather than extrapolating aggregated figures of merit (macrovariables). For example, aeronautical engineer N. S. Norway's forecast of the future potentialities of commercial aviation probably failed for this reason: he predicted (1929) that by 1980 cruising speeds would be 110 to 130 mph, range 600 miles, and payloads 4 tons out of 20 tons gross weight for commercial aircraft [11]. This pitfall has been a particular bugaboo of engineers. Herman Kahn has pointed out that the Scientific Advisory Board of the Air Force and the physicists at Los Alamos were comparatively poor at predicting the future of nuclear weapons technology, probably because they had too much "expertise" to look at the whole picture. Forecasters at RAND Corporation, on the other hand, predicted more accurately by "naïvely" extrapolating from envelope curves [12].

5. *Incorrect calculation.* The ironical case of Simon Newcomb and his ill-timed debunking of the airplane is well known and has already been cited [13]. Another famous example is the calculation by the Canadian astronomer, J. W. Campbell, that a moon rocket would have to weigh 10^6 tons in order to carry 1 lb of payload [14] (he was off by six orders of magnitude, due to unrealistic assumptions about fuels and failure to take multiple staging into account). Another interesting misconception was initiated by J. B. S. Haldane [15] and has been propagated by Dennis Gabor in his well-known book *Inventing the Future* in connection

with the question of future means of feeding the world's population [16]. Haldane's prediction was that some new artificially bred species of nitrogen-fixing algae would vastly multiply the capacity of the sea to sustain life. This overlooks the fact (known to contemporary biological oceanographers) that the amount of protoplasm in the sea is at least as limited by the availability of phosphorus and iron (as by nitrogen) in the surface waters—a constraint that is unlikely to change since there is no phosphorus in the atmosphere which can be "fixed" [17].

6. *Intrinsic uncertainties and historical accidents.* In addition to the foregoing pitfalls, it must be recognized that technological progress often depends in some degree on basically unpredictable elements such as luck or coincidence, individual insights, and quirks of personality. Certainly there are many examples in history of situations where a small accidental event leads to a large difference in the outcome. Thus, "for want of a nail the shoe was lost," etc. There is a considerable amount of speculative literature based on "What if . . . ?" propositions; e.g., What if Richard III

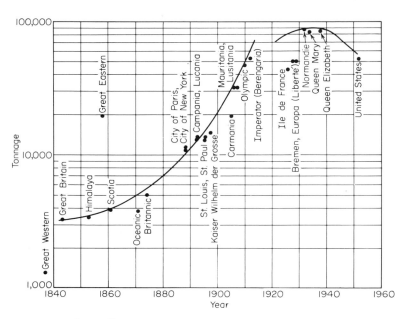

figure 2.6   *Ocean liner tonnage.*

hadn't been unhorsed at Bosworth Field? What if John Wilkes Booth's pistol had misfired? The history of technology also has its share of obvious (or not so obvious) examples. For instance, suppose that the discovery of the diffraction of electron beams had occurred *before* Planck's explanation of the black-body radiation spectrum and Einstein's consequent search for, and discovery of, a photoelectric effect. If the wavelike nature of particles had thus been found before the corpuscular nature of electromagnetic waves (rather than vice versa), quantum mechanics might have been invented almost immediately by a simple extension of the electromagnetic theory of James Clerk Maxwell.[3] The intellectual agony which afflicted theoretical physics in the 1920s, resulting from an apparent contradiction in nature, might have been avoided if contradiction had not been noticed until after its resolution had already been found. The central line of development of modern physics might have been considerably different, therefore, if two simple experiments, neither of which depended on the other, had occurred in a different sequence.

There are numerous other "What if . . . ?" examples which could be cited to prove, if proof were necessary, that luck, coincidence, and "human factors" make prophecy a very chancy business. What if Dr. Alexander Fleming or one of his colleagues had had the entrepreneurial propensities of a Dr. Squibb and had pioneered the commercial development of penicillin himself, instead of waiting to be discovered by the Rockefeller Foundation? What if Hermann Ganswindt, who "flew" by a helicopter of his own design in 1901, had been a better engineer and a less fanatic martyr?[4] What if Kammerlingh-Onnes, who first liquefied helium

[3] In fact, the time-dependent Schrödinger equation is an obvious extension of Maxwell's wave equation. On the other hand, the theory of special relativity would have had to provide Dirac's one-electron equations from the start, and would have had to face up to apparent contradictions, some of which are still not fully resolved [18].

[4] The Ganswindt helicopter had no engine because none powerful enough then existed and only flew vertically on a pole, powered by a torque supplied by a weight dropping into a well. The design had zero stability and would have crashed in free flight [19]. On the other hand, these faults could have been corrected within a few years.

in 1908 and discovered superconductivity in 1911, had bothered to carry his experiments just a little further and noted the Meissner effect and the phenomenon of superfluidity (which were not in fact found until 1933 and 1938 respectively)? Or what if Sir James Dewar, upon hearing of Kammerlingh-Onnes's reported "success" in 1908, had not become discouraged and ceased to pursue his own parallel researches? Finally, what if an efficient mono-tube steam "flash-boiler" had been developed prior to Charles Kettering's self-starter, rather than a few years later? By the time a quick-starting steam car was built (the Doble, produced in small numbers from 1922 to 1930), the mass-produced internal-combustion engine had become too solidly entrenched to challenge.

REFERENCES

1. Technical Bulletin No. 2, U.S. Navy, Bureau of Ships, January, 1941. This example was cited by D. G. Samaras in a lecture entitled "The Portents of the Nuclear Space-Age," delivered to the Satellite Seminar of the NCS-AIAA, Jan. 22, 1964.
2. A. C. Clarke, *Profiles of the Future: An Enquiry into the Limit of the Possible,* Victor Gollancz, Ltd., London, 1962.
3. N. Talensky, *International Affairs,* p. 16, October, 1964.
4. G. H. Stine, "Science and Fiction Is Too Conservative," *Analog: Science Fiction, Science Fact,* May, 1961.
5. D. G. Samaras, "Nuclear Space Propulsion: A Historic Necessity," *Nuclear Energy,* p. 352, September, 1962.
6. Figure by author.
7. G. H. Stine, *op. cit.*
8. J. F. Hogerton et al., *Atomic Energy Deskbook,* Reinhold Book Corporation, New York, 1963.
9. Chicago Area Transportation Study, Final Report, vol. III, April, 1962.
10. S. C. Gilfillan, "The Size of Future Liners," *Independent,* vol. 74, pp. 541–543, 1913. Cited in S. C. Gilfillan, "A Sociologist Looks at Social Progress," in James Bright (ed.), *Proceedings of the 1st Annual Technology and Management Conference,* Prentice-Hall, Inc., Englewood Cliffs, N.J., 1968.
11. Clarke, *op. cit.*
12. Herman Kahn, personal communication.
13: See Clarke, *op. cit.*
14. J. W. Campbell, "Rocket Flight to the Moon," *Philosophical Magazine,* January, 1941 (cited by Clarke).
15. J. B. S. Haldane, *Daedalus, or Science and the Future,* Kegan Paul, London, 1924.

16. Dennis, Gabor, *Inventing the Future,* Alfred A. Knopf, Inc., New York, 1964.
17. For instance, K. Brandt, "Beiträge zur Kenntnis des Chemischem Zusammensetzung des Plankton:" *Meeresunters. Abt. Kiel,* vol. 3, pp. 43–90, 1898. Or, see A. C. Redfield, "Biological Control of Chemical Factors in the Environment," *American Scientist,* September, 1958.
18. Nandor Balasz, personal communication.
19. See Willy Ley, *Rockets, Missiles and Space Travel,* The Viking Press, Inc., New York, 1951.

3 EPISTEMOLOGY OF FORECASTING

TECHNOLOGICAL CHANGE

There are two fundamentally opposed ways of looking at the dynamics of technological change, which may be termed the "ontological" and the "teleological"[1] viewpoints. As with other phenomena having inherently contradictory alternative interpretations[2] the truth certainly lies somewhere between, although it by no means follows that the two poles of the duality should be given equal weight. However, a synthesis of some kind may be appropriate.

The first or *ontological* view is that invention and innovation are visible manifestations of a self-generating process or an institution having a dynamism and a life of its own. Having once been

---

[1] The word teleological is intended to convey the notion of (social) purpose, but not an Aristotelian "final cause" or direction by the Deity. In the literature the word "normative" is often used in this sense. However the latter term implies guidance by definite rule or law, which is too strong.

[2] For instance, "heredity versus environment."

29

initiated, by whatever complex of prior causes rooted in history and culture, the subsequent growth of science and technology must now be understood primarily in terms of response to scientific or technological opportunities or challenges. These may be classed as *endogenous* or *intrinsic* variables.

A number of physical scientists hold this view, although it does not normally represent a self-conscious and explicitly defended position. Examples can be found in the writing of C. C. Furnas [1] and Gerald Holton [2], whose "model" of scientific progress is described in Chap. 7. Many members of the general public also seem to share this attitude, often viewing science and technology (with alarm) as if these were an uncontrollable "sorcerer's apprentice." Indeed, the common attitude that the dangerous or socially undesirable consequences of technology (particularly in the field of weaponry) are somehow the responsibility of science as an institution and, by extension, of scientists as individuals, implies that science and technology have not only a life but almost a will of their own [3]. The attitude typical of many political and social historians that inventions with important consequence are themselves random or "uncaused"—or, at any rate, determined by "hidden" variables out of reach of the central motive processes of history—is, again, characteristic of the view of science as a "force" in itself. Until recently the standard attitude of economists was similar: technological change was viewed as an exogenous variable (by implication) beyond the control of the marketplace.

The contrasting *teleological* view is, of course, that invention and especially innovation are actually impersonal social processes determined by social or military needs or by the existence of an effective economic demand. The importance of the individual inventor is discounted; if Edison had not invented the electric light—so goes the argument—someone else very soon would have done so, because "the time" of electricity had arrived. This theory seems to account not only for the remarkable number of cases of nearly simultaneous independent inventions or discoveries[3] but also for the cases of premature inventions—like Babbage's calcula-

[3] As tabulated, for instance, by Ogburn and Thomas (1922) [4] or Bernhard Stern (1927) [5].

tor or Stirling's hot-air engine—which failed, only to be revived later when social needs demanded them.   Unfortunately, although the observed frequency of simultaneous invention has been cited by sociologists such as Ogburn and Gilfillan in support of the hypothesis that technological progress is controlled by external, social forces, the evidence cannot legitimately be stretched so far.   The most that can be inferred is that invention tends to occur in response to a stimulus or signal of some kind, which might, however, equally well be an endogenous event such as a widely perceived opportunity or scientific gap.   In fact one of the most famous examples of coincidence—the simultaneous, independent invention of differential calculus by Newton in England and Leibnitz in Germany—could hardly be explained in terms of the social or economic needs of the time.   On the other hand, this may well have been the logically inevitable next step in the natural, evolutionary development of mathematics, and therefore in some sense "due."   However, notwithstanding these cautionary remarks, one can argue that, on the whole (but not always), the practical application of inventions occurs only when there is a ready-made external demand of some kind.   In other words the market precedes the product more often than vice versa.

The primacy of external (societal) influences would suggest that technological change can be forecast, if at all, only insofar as it is a consequence of (i.e., a response to) altered demands or requirements imposed from outside the research and development "system."   Thus if a social need is recognized, a technological solution to it might be inferred.   Gilfillan formalized this approach into the principle of "equivalent invention."   Thus in an essay for *Technological Trends and National Policy* in 1937, he counted 21 alternative hypothetical means of flying into the air, 6 ways to make uranium explosive, 18 methods of contraception, 25 methods of landing an aircraft in fog,[4] etc. [6].

Unfortunately, social needs are not *ipso facto* a reliable guidepost to technological change, because even widely recognized needs do not necessarily call forth invention or innovation; a need is not

---

[4] It is curious that, while nine of these methods were in use by 1952, the most important—radar—was not included in the 1937 list! [7]

always reflected in a market. Pollution, congestion, crime, over-population, malnutrition, poverty, and race relations have all proved to be comparatively impotent as stimuli to technology. In this they differ radically from other social needs, such as the demand for labor-saving machinery resulting from a shortage of farm labor in the United States during the nineteenth century,[6] or the demands for personal mobility, communication, electric power, and so forth which seem to stimulate innovation in these fields. The crucial difference is the existence of a suitable market mechanism in the latter case, while there is none in the former. Pollution, congestion, and so forth are so-called "externalities" or "third-party" effects; an individual cannot exercise a choice in the market-place as to the cleanliness of his air or water, for instance [8]. Only a consumer's association or a public agency can fill this role, and only when it does so will the generalized need for solutions to problems be translated into a specific demand for inventions and technological innovations.

The classical model for intervention by a public agency is, of course, the military establishment, which is the institutionalization of a need for national (and internal) security and stability. When military requirements become paramount, as during war, they have been historically the most effective of all stimuli for technology. Recently military demands for technology have continued even during times of (relative) peace.

Of the theories of intrinsic and extrinsic determinants of technological change, it would appear that the second is probably much the most important in the long run (with the caveat that it is not social needs per se which stimulate technology, but the translation of those needs into explicit demand in a marketplace, which may have to be created especially for the purpose). Wherever a competitive market exists, the technology which develops will reflect the economic interactions between buyers and sellers. Thus if automobile buyers want safety, *as individuals,* competition among

---

[5] Almost all farms outside the South relied entirely on large families (and animals) for labor; hence the immediate acceptance of mechanization when it became available.

entrepreneurs will ultimately provide safe cars; if they want high acceleration and maximum power, they will get them; if minimum fuel consumption were a primary consideration, it would be forth-coming (assuming competition exists among entrepreneurs). The determination of what the customers "really" want is in reality always a matter of uncertainty and is never final: success in antici-pating consumer preferences is a necessary but not a sufficient con-dition for success in the marketplace. Since the customers are many and their wants change slowly, it is often reasonable to fore-cast technological change *as though* the science and technology were subject to an internal opportunity-oriented law of develop-ment. This indicative procedure has been called *exploratory fore-casting,* and we shall retain the terminology.

However, if the "market" is a unitary one with only one effective customer (i.e., a government agency), there need be no such uncer-tainty about technological requirements.[6] In what might be termed a *unitary market* ultimate objectives can be (and are) spelled out quite precisely. The determination of exactly what technological changes are needed to achieve a specific mission, by means of a variety of intermediate steps or routes, is a more nearly deductive exercise, which has been called *normative forecasting* by Erich Jantsch [9], although teleological forecasting would be more accurate in most cases. This approach is a prerequisite for a truly detailed research and development plan or program leading to a prescribed objective via a prescribed path.

A normative or teleological forecast, as above defined, is only a statement of what ought to be (or needs to be) possible at some future time. To be useful in planning, something more is needed, namely an indication of feasibility. Such indications must, in prin-ciple, be derived from a set of development projections, parametri-cally indexed by level of investment or support. In practice, of course, the most reliable baseline is an extrapolation of the rate of invention given a continuation of the present level of support (or demand). Deviations from this baseline require a "scaling

---

[6] By the same token, there is less assurance in regard to *continuity* of requirements; goals may be changed suddenly and, indeed, often are.

law" of some sort.[7] A detailed discussion of these techniques is given elsewhere (Chap. 7).

Disregarding the rare situations where technological change is entirely "pulled" by explicitly defined objectives and missions in a unitary market—as is perhaps the case in the aerospace area[8]—it is fair to say that the existing research and development establishment often has some residual tendency to influence technological progress by "pushing" in the directions opened up by scientific discoveries. The "pull" of objectives usually outweighs the "push" of opportunities, but the latter is nevertheless important in many instances. In recent years xerography and lasers have been outstanding examples of areas where substantial development efforts were undertaken on the basis of beckoning opportunity rather than known requirements or recognized demand.[9] The problem of forecasting where both types of dynamism exist is clearly more complicated than where either one exists in isolation.

TYPES OF FORECASTS

In a general sense the "purposes" of forecasting closely match the contrasting views of technological change described previously. That is, one may ask (1) Where *is it possible* to go from here? What are the dangers or opportunities? (thesis) or (2) Where do we *intend* (or *desire*) to go from here? What are the goals? (antithesis). Finally, of course, one may attempt to answer the question (3) Where do we *expect* to go from here? What are the most probable paths? (synthesis). A forecaster may be concerned with any one of these or with all of them. However differ-

---

[7] The simplest such law would be linear, viz., double the investment, double the rate of invention; however, for science as a whole the invention rate seems to vary as a much smaller ($\sim \frac{1}{4}$) power of the financial support. A direct relationship between invention and investment is hard to deduce, but costs seem to vary as the square of research and development manpower [10].

[8] Nevertheless, prior to World War II technological progress in rocketry was largely opportunity-oriented. Teleological thinking was certainly present—Oberth, Goddard, Von Braun, et al., had a very clear ultimate objective (space travel)—but there was no market whatsoever until the German military establishment provided one (for different purposes).

[9] In fact lasers have repeatedly been called "solutions in search of a problem," which is another way of saying the same thing.

ent attitudes as well as different methods are appropriate to the different basic purposes, and it is useful to try to make these distinctions more explicit.

EXPLORATORY PROJECTIONS (POSSIBLE FUTURES)

Under this heading belong a wide variety of "What if . . . ?" questions, all of which might be lumped together as *chains of conjecture*. These can be distinguished primarily on the basis of the nature of the supposition or the "if" in the question. Thus the simplest subclass of exploratory forecasts is the *trend extrapolation*.[10] Here the underlying supposition is simply that the "environment"—or the balance of forces—does not change, so that it is reasonable to assume that the behavior of the recent past is a good model for the behavior of the near-term future.

One of the principal uses of simple extrapolation is to identify major future problems (and, by alerting attention to them, help to ensure that they will be solved). Examples are commonplace. The problem of overpopulation and the threat of mass starvation in underdeveloped countries is one which has been repeatedly brought to light by simply comparing extrapolations of population and food production. It could have been foreseen a generation earlier if the full impact of modern medical advances on death rates had been adequately considered by demographers. The problem of building houses and schools for the "baby boom" of the fifties was revealed by simple extrapolations. Similarly we are currently anticipating serious shortages of water, rapidly increasing crime rates and traffic congestion, and possibly dangerous climatic and ecological imbalances[11] by the year 2000. In the same vein one can project disastrous problems of radioactive and thermal pollution, increasing nuclear proliferation, and so forth.

There are a number of obvious variations of the method of extrapolation. For instance, one can ask what would happen,

[10] By a "trend" one can mean a number of different things, but in general terms we refer to a time series of some figure of merit.

[11] Caused by excessive CO_2 production and too little oxygen, due to rapid consumption of fossil fuel [11].

ceteris paribus, if one external condition were altered? The single change to be taken into account could be a policy proposal (What if the "massive retaliation" doctrine is scrapped? What if Roman Catholic objections to artificial means of birth control were changed?), an event (e.g., massive crop failure in India), a change in research budget level, or a technological breakthrough with a great impact on a major industry. Projections of the latter type have been particularly numerous, perhaps because they have so much intrinsic interest.[12] Many useful "scenarios" of space travel, automation, psychic manipulation, and so forth have been generated in this manner.

When the "if" in the "What if . . . ?" question being asked involves changing several variables simultaneously, or when there is reason to think several variables will change over the time range of the forecast, even qualified extrapolation ceases to be an appropriate tool for analysis.) One must start to worry about multiple correlations among variables; the situation becomes more complex, and it is less and less sensible to make meaningful projections without understanding the underlying social dynamics involved. From naïve and not-so-naïve extrapolations one is led by degrees to construct *models* and to make predictions from them.[13] This activity may be termed *"heuristic forecasting."* In principle, social models may involve any number of variables; however, to date, most forecasting models have concentrated on relating the rate of invention or innovation in a field to such factors as information flow, exhaustion of possibilities (or resources of "ideas"), research budget levels, manpower, and the like (see Chap. 7).

Whereas the simple extrapolation is primarily used to reveal problems which require urgent attention, qualified extrapolations *ceteris paribus* or the use of social models to generate scenarios are mainly beneficial to decision makers who are required to opti-

---

[12] Fiction writers often go further and explore the consequences of "impossible" scientific breakthroughs such as antigravity, time travel, ESP, immortality, or speeds greater than the velocity of light. Some of these may ultimately prove to be not impossible after all. Indeed, at the present time the absoluteness of the speed of light is being seriously questioned by some physicists [12].

[13] Daniel Bell calls this "social physics" [13].

mize a policy or a "system"—transportation, communications, electric power, strategic weapons, etc.   The systematic evaluation of a proposed policy or systems concept against many alternative "futures" is an increasingly standard part of government and industry planning.   In everyday language this means estimating the performance of a piece of equipment or a contingency plan in a wide range of circumstances.   A rational choice of program objectives and mission requirements cannot logically precede such an evaluation.

Obviously many pertinent alternative futures do not involve significant technological changes, whence scenarios can be generated without technological forecasting; but this happy situation is becoming increasingly rare.   For instance, a significant aspect of the future environment of a weapon system depends on the other weapons which will become available—particularly to the opposition.   Similarly in industry, sales potential of a new technological product depends vitally on the competition (both in the marketplace and on the technological front).

A closely related application of technological alternative futures is in making *benefit-cost* calculations to guide investment in large-scale capital projects.   It is clear that the projected benefit from a highway or a dam or a pump-storage plant depends not only on the anticipated demand for the functional capabilities being provided (transportation, water, or peak power) but also on the technological alternatives which will exist during the time frame over which the calculation is extended.[14]   Thus the future benefits in question cannot be evaluated without also estimating the prospects for air and high-speed rail transportation; desalting plants and artificial rain making; and for gas-fired turbogenerators and superconducting-ring storage devices.   Failure to consider the technological alternatives may result in serious misallocations of resources.

A different and, in some ways, more sophisticated exploratory forecast results if we do not stress the continuity of the future with the present but, instead, focus on the discontinuities and qualitative changes.   In the realm of invention and technology, this ap-

---

[14] E.g., 50 years for water resources projects.

proach has been formalized and called the *morphological method*.[15] Whereas an extrapolation assumes a series of incremental changes in a certain known direction, the morphological approach has nothing to say about the rate or direction of invention but systematically catalogs all possible opportunities for invention in a field.   In the hands of a skilled analyst the approach permits systematic exclusion of unworkable combinations and comparisons of generalized performance characteristics of the survivors.   Thus it is a tool for the guidance of inventors.

It can be used also, however, as a basis for assigning a priori probabilities for the direction of technological evolution, especially of large composite systems.   The rationale for this usage is discussed extensively in Chap. 5.

TARGET PROJECTIONS

These are immensely important in practice, although they are not strictly forecasts at all.   One crude type is essentially advertising, designed to stimulate interest and action.   Writings of aviation and rocket pioneers, modern architects, utopian city planners, and some others seem to fall into this category: to the extent that they are prophetic they are, in part, self-fulfilling prophecies.[16]

Target projections are often used as tools for policy makers. They may serve either of two roughly distinguishable functions: (1) as a detailed guideline or measure of achieved versus hoped-for progress in the area with which the projection is concerned or (2) as a basis for making decisions on matters extraneous to the projection itself.   To illustrate the first, economic growth is desired for its own sake and the actual performance of the economy is constantly being measured against stated goals.   Deviations (at least on the downside) sometimes spark compensatory monetary and/or fiscal policy changes.   A good example of (2) is population growth, which is not, nowadays, desired in itself but which must

[15] The morphological method is not confined to exploratory forecasting of opportunities; one can similarly analyze functional capabilities, tasks, missions needs, or goals, as will be seen.

[16] The writings of Marx and Engels are undoubtedly the best examples of this phenomenon.

be assumed to make current decisions about investments in school construction, for instance.

These cases do not involve technological change, but it is not difficult to find examples which do. Thus an electronic computer manufacturer may make a 10-year target projection for future information-processing capability per dollar; or the FPC may project future power costs. These are directly related to socioeconomic goals, and the annual progress of the company or industry may be gauged by comparing its achievements with such projections. On the other hand, projections of these—or of other—figures of merit may be used as inputs for quite independent questions such as designing future weapons systems or deciding whether or not to invest in developing an electric car.

Methodology is not as important in making target projections per se as it is in exploratory or heuristic forecasts. To be sure, technical feasibility is one factor to be taken into account in choosing performance goals, but in most instances it will not be a limiting one.[17] Anticipated costs, markets, compatibility with overall corporate goals, cross-fertilization, "image," and many other factors are involved in the choice.   /

Indeed, elaborate "decision trees" and other models have evolved to assist planners in sorting out all these complex issues. In fact the morphological approach mentioned earlier is also pertinent at other levels of the hierarchy: one can analyze the universe of all possible functional capabilities, all possible applications (missions, tasks), all possible strategies, all possible purposes (goals), and all possible basic attitudes or views of life. Having done so, one can (in theory) evaluate a given invention or system in terms of its overall relevance to the elements in this overall "demand spectrum." An invention or innovation may be relevant only to a single functional capability, which in turn has only one or a few potential applications. These, in turn, may only fit into a bizarre strategy to achieve an unlikely purpose. An instance might be a device to utilize the thermal gradient between the light (hot) and dark (cold) side of the moon to generate eletric power which

---

[17] The major exception is the military area, where the target projection is not infrequently specified as the maximum feasible performance projection.

could be used to power a vehicle designed to follow the moving shadow. If so, the invention is clearly not in demand. On the other hand, an invention may be relevant to many capabilities, many applications, many strategies, and many purposes. Clearly the most important inventions—e.g., the transistor—are of this character. Decision trees, relevance trees, and other planning models are discussed in Chap. 9.

VALIDITY CRITERIA

We have hitherto explored distinctions in forecasting arising from differing interpretations of the underlying social dynamics of technological change and from the various purposes of the forecaster. A third classification might be derived from cognitive considerations or "levels of understanding."

All forecasting is, *ipso facto,* an attempt to infer future events on the basis of what has occurred in the past (our only source of data). In this general sense, therefore, all forecasting methods are based upon the analysis of trends. However, it is cogent to remark that projection is simply one type of inductive inference: construction of a theory to predict the results of a hypothetical scientific experiment on the basis of data from previous experiments is a common application of the same principles. Our explicit concern with "the future" is largely incidental, as far as basic methodology is concerned.

It is useful to classify the various sorts of inference in terms of the degree of confidence one puts in them. One might distinguish the following hierarchy of cognitive levels, starting with the lowest:

1. Conjecture: probable positive correlation between some pairs of observations, as, for example, weather and sunspots.

2. Metaphor or analogy: the outlines of a coherent (but perhaps misleading) pattern are perceived, e.g., Plato's cave, or the analogy of birth, growth, maturity, and death to describe the life cycle of a technology.

3. Quasi model: more than (2), less than (4). Qualitative or operational predictions can be tested: examples include Darwin's

theory of "survival of the fittest," or the dictum "Ontogeny recapitulates phylogeny."

4. Empirical-phenomenological model: the pattern is adequately predicted by a mathematical formalism with empirically fitted parameters. Examples include "Hooke's law" of elasticity, "Ohm's law" relating voltage or current and electrical resistance, Landau's "two-fluid model" of quantum superfluids.

5. Analytic model: the pattern of events can be predicted and explained in terms of more fundamental "laws," with wide applicability, such as Maxwell's electromagnetic theory, Einstein's special and general theories of relativity, and Dirac's formulation of quantum mechanics.

Cognition, of course, is used in the sense of *understanding* or *explanation*. It has been asserted by Hempel [14] that explanation and prediction are symmetrical processes—differing essentially in regard to direction in time—both being deduced from a "covering law," as expounded by Karl Popper [15]. As Suchting points out, the symmetry is imperfect since an explanation must, by definition, be "true" whereas a prediction may be right for the wrong reasons, or may be arrived at without an explanation [16]. Indeed a simple trend extrapolation does not presuppose any understanding of underlying causes: it is enough that the (hidden) causes be unchanging in time. Only when changes occur in the balance of forces does it become necessary to worry about explanations and, in response to this concern, to generate models and theories which permit more sophisticated types of projections and models.

The choice of a level of inference is also obviously dependent on the reliability and completeness of one's data base (concerning past performance of "the system") and on the type of information required. As to this latter point, it is interesting to consider an ordered list of possible questions, ranked roughly according to the amount of knowledge required to answer them:

1. Will event X occur (if we assume A)?
2. Will event X precede the year 1980?
3. Will event X precede event Y?
4. When will X occur?

5. If X occurs in 1975, when will Y occur?

6. When will X and Y occur?

It is understood in all cases that it is only reasonable to seek something on the order of a 90 percent confidence level. For question 1 (depending, of course, on X and A) very little understanding of underlying mechanisms may be required; indeed questions 1 through 4 may be answerable by simply extrapolating trends. However, question 3 requires at least a minimum of knowledge about the relation between X and Y, while question 4 requires some knowledge about the relation between many (unspecified) exogenous variables on X. Questions 5 and 6 require even more detailed knowledge about the relationship between X, Y, and the environment.

The desirability of being able to answer questions of the latter type is not in doubt, but many of the questions about the future which are of greatest practical concern are nevertheless of the simpler type 2 or 3. For example, one might ask, "Will the world be able to feed its population in the year 2000 at the same level as today?" Since the answer depends on whether population or food production rises faster, a nearly equivalent question can be rephrased as follows: Which will double first, population (event X) or food production (event Y)?

Short-term questions are often of type 4, viz., when will the Dow-Jones average hit 1,000? When will the gross national product reach a trillion-dollar annual rate? When will the United States population pass 200 million? These are all essentially equivalent to asking for rates of increase of the relevant variables.

While questions 2 and 3 may be answered in yes-no terms, questions 4, 5, and 6 are more demanding and quantitative. This means that a phenomenological or empirical model is required in the latter case, whereas a less sophisticated method may be adequate in the former. Naturally, most long-range questions of practical significance will usually involve combinations of various levels of inference. Indeed one may proceed by several parallel chains of reasoning, using different conjectures, analogies, and "covering laws," only to arrive ultimately at contradictory conclusions. Deeper explanations may be needed to resolve the discrepancies.

As an illustration of this point, let us consider briefly the question previously raised: population versus food production in the year 2000. One can argue, as some have, that the average rate of population growth is likely to increase, because technological progress in the field of medicine will cut the death rate while the birth rate tends to remain nearly constant (if we extrapolate present trends). The rate of increase of food production, however, might go down simply because the available resources are subject to natural limits (acreages, rainfall, photosynthetic efficiency of plants, etc.), and as agriculture approaches closer to its theoretical maximum efficiency, each increment of increased production requires greater and greater efforts to achieve.[18]

On the other hand, one can also argue that, in principle, food production need not be limited by photosynthesis at all. There are other theoretical means of synthesizing proteins, e.g., by fermentation of hydrocarbons (which can, in turn, be synthesized from elementary carbon and hydrogen, with the addition of energy). It is a question basically of the cost of energy relative to the accumulated capital and knowledge which exist. Hence, the counterargument might go: if society can accumulate nonagricultural wealth rapidly enough by exploiting cheap resources, it will become possible to synthesize food at a later time relatively cheaply *in terms of the wealth then existing.*

Another highly controversial subject in which contradictory forecasts can be (and have been) made relates to city planning in the year 2000. Shall we have to allow for 250 million private automobiles? Or 150 million? Or 100 million? A naïve extrapolation of present growth rates suggests a number something like the first. But there are several converging factors which might lead to an earlier topping-out of the curve. Traffic congestion appears to be an insoluble problem; continuing urbanization coupled with skyrocketing urban insurance rates and garaging costs may accelerate the tendency to switch from a pattern of individual ownership to a pattern of rental (by the hour, day, month, or

[18] This argument is hardly watertight as it stands; to defend it seriously one would have to add a series of subsidiary qualifying statements. However, at this stage we are not defending or attacking, but merely illustrating a possible line of reasoning.

year). Computer-dispatched door-to-door jitney service may provide an attractive form of public transportation which would eliminate the need for a second car in the suburbs, etc. Again, a naïve extrapolation is risky: a deeper understanding of the underlying dynamics of change is crucial in this case.

These two examples suggest that a decision which must often be faced by practical forecasters in any given case is: can one rely on simple extrapolation (disregarding mechanisms) or not? In making a yes or no decision, it is often as important to know that a particular variable is *not* involved as to know definitely that it is, if there is reason to expect a change in that variable. For instance, it would normally be possible to state unequivocally that X does *not* depend appreciably on the price of silver, whence a change in the price would not be a salient factor and *ceteris paribus* a naïve extrapolation would be satisfactory. Indeed one can not infrequently make even stronger statements, such as that X is independent of changes in any or all of the environmental factors which one expects to vary significantly during the time frame of interest. On the other hand, if the nonrelevance of one or more variables to X *cannot* be established, the situation becomes much more complicated, simple extrapolations are unreliable, and a higher level of understanding (cognition) is needed.

REFERENCES

1. C. C. Furnas, *The Next Hundred Years: The Unfinished Business of Science,* The Williams & Wilkins Company, Baltimore, 1936.
2. Gerald Holton, "Scientific Research and Scholarship: Notes toward the Design of Proper Scales," *Daedalus,* March, 1962.
3. See, for instance, Jacques Barzun, *Science: The Glorious Entertainment,* Harper and Row, Publishers, Incorporated, New York, 1964.
4. William Ogburn and Dorothy Thomas, "Are Inventions Inevitable? A Note on Social Evolution," *Political Science Quarterly,* vol. 37, no. 1, pp. 83–98, 1922.
5. Bernhard Stern, "Social Factors in Medical Progress," 1927. Cited by S. C. Gilfillan, *Technical Trends and National Policy,* U.S. National Research Council, 1937.
6. S. C. Gilfillan, *Sociology of Invention,* Follett Publishing Company, Chicago, 1935.
7. S. C. Gilfillan, "The Prediction of Technical Change," *Review of Economics and Statistics,* vol. 34, pp. 368–385, 1952.

8. R. U. Ayres and A. V. Kneese, "Product Consumption and Externalities," *American Economic Review* (in press, 1969).
9. Erich Jantsch, *Technological Forecasting in Perspective,* OECD, Paris, 1967.
10. See William Farrington, "Squeezing Research Dollars," *International Science and Technology,* September, 1963.
11. *Restoring the Quality of Our Environment,* Appendix Y.4, Report of the Environmental Pollution Panel, President's Science Advisory Committee, Washington, D.C., 1965.
12. See G. Feinberg, *Physical Review,* vol. 159, p. 1089, July, 1965.
13. Daniel Bell, "Twelve Modes of Prediction: A Preliminary Sorting of Approaches in the Social Sciences," *Daedalus,* Summer, 1964.
14. Carl Hempel, *Aspects of Scientific Explanation and Other Essays in the Philosophy of Science,* The Free Press of Glencoe, New York, 1965. Cited by Dennis Johnston in an excellent unpublished paper "Long Range Projections of Labor Force" (preliminary), U.S. Bureau of Labor Statistics, 1967.
15. Karl Popper, *Logik der Forschung,* 1935. Cited by Johnston, *op. cit.*
16. W. A. Suchting, "Deductive Explanation and Prediction Revisited," *Philosophy of Science,* vol. 34, March, 1967. Cited by Johnston, *op. cit.*

4 CONTEXT: DIMENSIONS OF TECHNOLOGICAL CHANGE

In the previous chapter we examined the various forms and aspects of forecasting technological change as an intellectual activity. Now we must focus on the subject matter of this dialectic: What exactly is meant by technology? What is technological change? How does technological change relate to other human functions?

A pitfall for forecasters which might well have been included in Chap. 2 might be the stricture: "too narrow a context." There are many kinds of issues "in the large" which influence technology and should, therefore, be considered in looking into the future. These include theological and philosophical attitudes, cultural background and culturally determined modes of thought, religious doctrine, political and international questions, military and strategic postures, and socioeconomic issues and policies.

Technology, as used hereafter, is the systematic application of organized knowledge to practical activities, especially productive ones. In the following it will occasionally be useful to have emphasized the distinction (at least once) between *invention, innovation, transfer,* and *penetration or diffusion.* The first refers to the birth of an idea for an artifact, process, or procedure, with some claim to novelty and priority, such as might be recognized by the U.S. Patent Office.[1] All technological change rests upon invention—not necessarily discovery—but only a small fraction of inventions actually have any application in technology. When such an application is formulated, resulting in new products, approaches or ways of doing or making things, an *innovation* has taken place. However, again, there are many more innovations than enduring changes. When an innovation proves successful and begins to displace the existing products or methods—as, for instance, the basic oxygen process has established itself in the steel industry—the phenomenon is called *penetration* or *diffusion.* The term *technology transfer* has been used in various ways by other authors, but it will be used here in the most common sense, namely, with reference to situations where an established technology in one field finds an unexpected application in some quite different area.

To illustrate the above definitions: the first germanium transistor was an *invention;* the first use of transistors in place of vacuum tube diodes in electronic computers was an *innovation* in computer design; the rapid replacement of vacuum tubes in this and other electronics applications including radio and TV would be described as *diffusion;* the use of transistors in applications where vacuum tubes were never used, such as alternators or invertors—where they replace relays or commutators—might be called *transfer.* Other classic cases of technological transfer have been documented in attempts to assess the impact of space research on non-space-related activities: the rapidly increasing use of electronic telemetering devices in medicine or the development of totally new approaches to living and working under water have been cited as pertinent examples.

In recent decades the phenomena of innovation, diffusion, and transfer have become progressively less haphazard and more subject to conscious management and manipulation. "Research" became

[1] Not all patentable ideas are really inventions, nor are all inventions patentable (especially in the realm of processes and procedures), but there is a significant degree of overlap between inventions and patentability.

divided into "research and development" and then "research, development, testing, and evaluation" or RDT&E. In the Defense Department, the development phase has also been split into "exploratory development," "advanced development," and "engineering development," which are followed by testing, evaluation, procurement, deployment, training, etc. Of course the explicit recognition and naming of these phases has not made the overall processes of innovation and diffusion any more complex or time consuming than they ever were, but has probably made them somewhat easier to administer, and possibly to analyze.

This proclivity to divide the process into more and more distinct parts also underlines the importance, in the following, of talking about progress of technology in time only with reference to a specified phase of development. In most instances it will be appropriate to consider either technology available in the *development phase* at a given point in time, or else technology available *commercially*. However, progress in time must be distinguished from progress from one phase to the next: in fact, it is convenient to think of the first as *horizontal* movement and the second as *vertical* movement in a 2-dimensional "transfer space."[2]

While this book concentrates mainly on the methodology of forecasting in the horizontal direction—which is, after all, what most people think of when they think of technological "progress"—we cannot ignore the vertical aspect of the problem. Indeed many long-range planners in government and industry are far more concerned with the latter than with the former: to some, the really significant question is how the industrial or governmental administrator can foresee (and expedite) the transition of research projects from the exploratory development to the prototype to the engineering-design and marketing phases. While this is primarily a function of organizational structure and overall corporate management philosophy, there are certainly nondeterministic aspects to the process, and it is therefore appropriate to apply a forecasting approach to some features of it.

In the last three centuries there has been a marked telescoping

[2] The concept of a technology "transfer space" is used differently by Jantsch [1].

of the time required for an important invention to reach maturity and full utilization. This trend, which is certainly due in large part to rapid improvements in the techniques of communications and social feedback, may itself be reaching a stage where random factors become unimportant and the process can itself be both managed and forecast. We shall return to this topic at the end of the present chapter. In the following, a number of external factors which may have a major influence on the rate and direction of technological change are discussed successively.

THE INTELLECTUAL-PHILOSOPHICAL-CULTURAL FACTORS

The range and scope of this category is enormous and mostly, like an iceberg, subsurface. At various times in history, in other cultures or even in certain Western "subcultures," attitudes toward science and technology have ranged from incomprehension and rejection through dogmatic orthodoxy, Romantic or Faustian, to the instrumental attitude which happens to characterize the leadership community of the West today. While we may feel justified in assuming that the latter point of view will continue to be dominant in the foreseeable future, we must also be aware that attitudes can change and other modes of thought coexist in the world. Thus, the "beat generation" and more recently, the "hippies," as well as Bible Belt fundamentalists, flat-earthers, Mennonites, Jehovah's Witnesses, Christian Scientists, and many other religious mystics or orthodoxies do not accept some significant part of the values and/or norms of the mainstream. Many such groups in the world today have doctrines or dogmas which are culturally—if not logically or theologically—inconsistent with the positivist, pragmatic orientation of modern science, just as the attitudes of the Medieval Schoolmen could not be reconciled comfortably with Roger Bacon or Galileo. There are also deviants in the other direction, of course, who—like the legendary Dr. Faustus, the Biblical Eve, or some present day Marxists (especially in Asia)—become fascinated with technical knowledge and progress for its own sake, and who believe that science is somehow synonymous with power or virtue or both.

A few years ago the dominant modes of thought seemed so solidly entrenched that it was difficult for most people to envision any change in the foreseeable future, unless brought about by some cataclysmic event such as a nuclear war.[3]   However, the rapid spread of an escapist "drop-out" philosophy among the young in recent years certainly suggests that Western scientific materialism may not, after all, be the inevitable "wave of the future."   Increasing material prosperity in the United States and in the other industrialized nations has been accompanied by disquieting signs of cultural disorientation and social malaise among both prosperous and deprived groups.   Rejection of the Western "Weltanschauung" may also be a fundamental driving force behind some of the current social unrest in the less developed areas of the world.   It is not impossible—some would say not unlikely—that some new and potent religious or moral-philosophical movement may make its debut on the world scene in the next few decades to fill a vacuum which has apparently been left by the decay of traditional religious beliefs.   Similarly, nationalism, communism, fascism, and other hitherto dominant "isms" appear to be declining in their ability to command the loyalties of men.   If this perception is valid, then new creeds less conducive to the fostering of scientific research as we know it may be in the offing.   By the same token, of course, it may be that science itself is to be the new religion of the world. An acute observer may discern evidences of either trend.   The important point is, of course, that the present intellectual environment is not necessarily invariant and immutable.

POLITICAL-INTERNATIONAL FACTORS

The importance of political and international problems (and resulting military strategic postures) to technological development is only too clear, since these are a major controlling factor behind current manpower allocations, research and development budgets, and pro-

---

[3] This is a standard theme of science fiction: a prominent example is *A Canticle for Liebowitz* [2].

grams.   Figure 4.1 illustrates the point by showing how the supply of scientific manpower was affected by World War II [3].   Figure 4.2 shows a similar effect due to the Civil War [4].   When the Vietnam War ends, we can anticipate a sharp decline in total military expenditure in subsequent years.   Although Defense Department expenditures on research and development will not, in all likelihood, fall correspondingly, there have already been (prior to 1965) widespread cutbacks in engineering industries and a reduction in "parasite" research and development programs funded "in-house" by major aerospace-industry contractors, suggesting that the rapid growth of research and development funding since World War II may be "topping out."   Indeed, there has been widespread concern in many quarters over the consequences of disarmament, or even a substantial further slowdown of military expenditure, to the current United States technological leadership of the world.

The impact of international competition on technology also occurs in ways other than through stimulation of military research and development.   The outstanding example is, of course, the race to put a man on the moon by 1970 (Project Apollo), which made NASA second only to the Defense Department in terms of disposition of research and development funds at its peak funding level in 1965.   The political nature of the entire space program is ob-

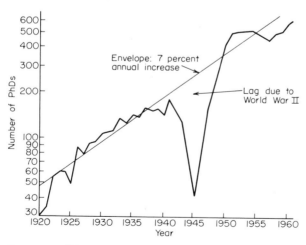

figure 4.1   *Physics Ph.D.s annually.*

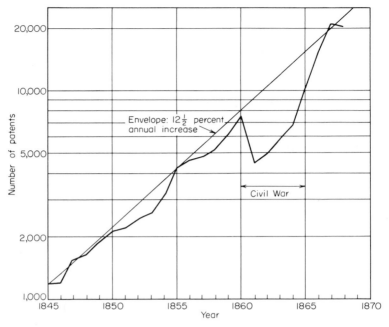

figure 4.2   *U.S. patents, Civil War period.*

vious.   The gradual loss of some of its original political sex appeal, after the first manned space flight, in turn caused NASA to concentrate considerable energies on identifying and supporting other civilian applications of space technology through its Office of Technology Utilization.   The success or failure of this deliberate attempt to generate technological spin-off may increasingly help to determine the success or failure of NASA in obtaining congressional approval of its budget requests.

Contemporary domestic political needs are increasingly drawing research funds into other areas.   Medical research was given a major boost by the formation of the National Institutes of Health. The Department of Health, Education and Welfare has been the fastest-growing in the federal establishment for more than a decade.   More recently still, there are signs of increasing political, public, and congressional interest in improving the quality of our environment.   Major programs have been started within the sixties on environmental health, air and water pollution, solid-waste disposal, pesticides, and highway beautification.   Transportation—

both in and between cities—is also receiving increased attention at the federal level. As soon as the Vietnam War comes to an end, all of these programs are likely to grow rapidly, resulting in a surge of new technology.

World population growth and the possibility of worldwide famine may trigger a new major cycle of federal-government-sponsored research and development within the next decade. Massive research programs to search for cheap, effective, and simple means of birth control for countries like India would doubtless have been put into effect already if it were not for domestic political opposition, mainly on the part of the Roman Catholic hierarchy. The Church itself is currently undergoing a rather glacial—but also painful—reappraisal of its position, and it is certainly not inconceivable that, some time after the present generation of Church leadership has retired from the scene, official dogma may change. Here, of course, the change would have to be as much cultural and sociological as political. Political support for major programs of research on nonagricultural sources of food—ocean farming, microbiological conversion of cellulose wastes or of petroleum— might come sooner. The political consensus is fragile, of course, inasmuch as important domestic agricultural interests might be (or might perceive themselves to be) adversely affected by such programs.

MILITARY AND STRATEGIC POSTURES

At the program level, military considerations have been of paramount importance to advanced technology for more than two decades, certainly since the Tizard Committee in England decided to give high priority to radar development in the mid thirties and since the United States decision to build the atomic bomb was taken, largely to pre-empt Hitler. Since the 1940s military planners have been increasingly obliged to hedge against possible technological breakthroughs by an opponent.[4] Defense Department

---

[4] We are using language loosely. More precisely, one *hedges* against what an opponent might (but probably won't) do and *designs* for the cases which seem most probable.

decision makers have often (and perhaps incorrectly) felt that prudence requires the assumption that anything which looks both useful and technically feasible will be developed by the adversary, and therefore should be developed by the United States. Under the McNamara regime the soundness of this argument was seriously questioned by the Pentagon, especially in regard to the issue of whether or not to build an anti-ballistic missile. Previously, the question "Is it technologically possible?" has sometimes been a major criterion in guiding research and development policy, especially in the Air Force. Hence, while prototype development is comparatively restricted in scope, much exploratory research and development supported by the military has been devoted simply to exploring the realm of the possible, with the hope of achieving militarily useful results as "by-products." This easy-going attitude is probably now a thing of the past as Project Hindsight—a careful evaluation of the effectiveness of military research and development programs [5]—seems to have confirmed the growing suspicion that non-mission-oriented research has been ineffective in terms of contributions to military programs.[5]

Military research in the mid 1960s changed course rather sharply away from heavy emphasis on the strategic (i.e., nuclear) threat toward greater emphasis on conventional and limited war. The nuclear weapons and ballistic missile programs, especially, have been downgraded somewhat in intensity. On the other hand the anti-ballistic-missile concept, which never convinced top Pentagon planners of its adequacy as an answer to the massive nuclear threat from the Soviet Union, has acquired a new impetus from the recent perception of a less overwhelming and less sophisticated threat from other quarters, viz. China.

In terms of specific areas of technology, it is obvious that the shifting winds of international alliances and hostilities have had major impacts. Thus the needs of the atomic energy program provided the first major stimulus to the electronic computer industry, and this impetus was maintained by the requirements of air defense (the SAGE system) and missile guidance. Electronic

[5] Military sponsorship of fundamental research can still be justified on other grounds, however: it is a major source of support for graduate education in the sciences, for instance.

data-processing technology would have lagged far behind its present level if it had not been for military support for research and development and military purchases of enormous numbers of computers, especially of the advanced types.

Similarly, military requirements for high-performance aircraft and, later, missiles, led to a great acceleration in aerospace technology. The civilian airline industry would not be the economic giant it is today if it still depended on planes like the DC-6 or the Constellation. Yet it is clear that the aircraft industry could not have developed the "707" and its competitors and successors without its prior experience with jet bombers. The British-French Concorde and the SST would as yet be undreamed of. The presently growing civilian market for helicopters also clearly has its genesis in military-sponsored development.

Obviously the civilian space program is also a direct outgrowth of the military development of ballistic missiles. Commercial nuclear power owes a substantial debt to the atomic bomb and the nuclear submarine. Recent progress in seismology—and hope for ultimate control of earthquakes—is largely due to support from Project VELA, the nuclear-test-detection program. Geodesy, cartography, geology, and meteorology have profited immensely from sophisticated surveillance techniques, especially long-range, high-resolution, infrared scanners in satellites, developed for military purposes. Advanced telemetry systems, originally needed for the missile and space programs, appear to have great potential value in medicine. Surgery has also profited greatly from techniques developed on the battlefield. Similarly, the long-standing military interest in undersea warfare has finally begun to show signs of paying dividends in terms of developing a capability for commercial exploitation of the mineral resources of the ocean floor.

It is crucial to note that these particular areas of important "spin-off" were not all inevitable consequences of military expenditure per se. That is to say, the areas where great and rapid progress actually took place might have been somewhat different if the perceived threat and the United States reaction to it had been different in detail. The nuclear submarine, for instance, might easily have come 5 to 10 years later than it did in fact—or not at all—if different men had been making the decisions. The ballistic missile program might equally have been initiated 5 or even 10 years sooner than it was, by a different set of decision makers; ditto the ABM program. The nuclear plane or the B-70 might have been continued, and so on. These alternative courses of action would have had different effects on the rates of technological development in various fields.

MACROECONOMICS

The effects of economic problems on technology are obviously of the first importance. An obvious example "in the large" is provided by the Great Depression of the 1930s. Figure 4.3 indicates that investment in farm mechanization virtually ceased (but recovered in the forties) [6]. This pattern was repeated in many industries. There are, of course, many subtler interactions between economic issues and technological development. For example, economic policies tending to decrease the labor supply will tend to have an impact on the potential market for automation; indeed, the rapid mechanization of United States agriculture, beginning

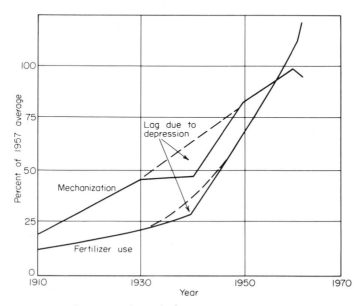

figure 4.3   *Investment in agriculture.*

in the nineteenth century, has been attributed in part to a continuous rural labor shortage.   The 1966–67 cutback of the "Bracero" program,[6] in order to make jobs available for higher-priced domestic labor, is likely to stimulate further farm mechanization instead.

Similarly, a shortage of supply or a rise in the price of some raw material or staple commodity is likely to stimulate the search for a technological alternative (and conversely).   The recent (1966) sharp world price increases for copper have resulted in a considerable acceleration in the shift of many former copper users, such as electric utilities and electrical appliance makers, to aluminum and other metals.   Thus aluminum is rapidly becoming the material of choice for high-voltage electrical transmission lines. Similarly, shortages of rubber, quinine, and other tropical products during World War II led to the development of substitutes.   Two-thirds of current United States rubber production is now synthetic.   The Germans, desperately short of petroleum, developed practical methods of producing aviation gasoline from coal (the Fischer-Tropsch process).   Other examples of substitution induced by scarcity are numerous.

It has become almost axiomatic, since World War II, that a dynamic technology creates new industries (electronic computers, plastics, atomic power, etc.).   The other side of the coin is that economic factors may contribute substantially to creating new technology.   Jacob Schmookler, an economist at the University of Minnesota, has assembled a good deal of data which suggests that technology, as measured, for instance, by the number of patents applied for (and later granted) in a field, year by year tends to have peaks and valleys closely coinciding with indices of the rate of domestic economic activity in the field [7].   This relationship apparently holds remarkably well both for long-term trends and shorter-term fluctuations.   The three charts (Figs. 4.4 to 4.6) illustrate the closeness of the correlation for two major United States industries, railroads and construction [8].   It may be objected, of course, that these statistical comparisons apply only to industries and periods where there was no government-sponsored research

_____

[6] An annual seasonal importation of low-wage labor from Mexico to harvest field crops (fruit and vegetables), mostly in California and Arizona.

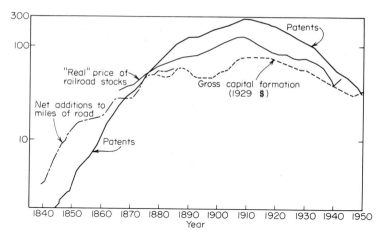

figure 4.4 *Railroad patents and railroad investment, United States, 17-year moving average. (Beginning with 1874 patents are counted as of the year of application. For earlier years they are counted as of the year of granting.)*

of any kind, whereas nowadays about two-third of our national research and development effort is directly financed by the government.

Nevertheless, in the private sector of the economy at least, there are persuasive indications that economic growth and technological innovation accompany one another. Certainly the obvious growth industries tend to be the most innovative and most research ori-

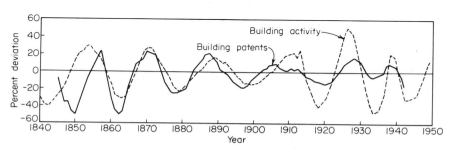

figure 4.5 *Building activity and building patents, United States, deviations of 7-year from 17-year moving average. (Patents counted as in Fig. 4.4.)*

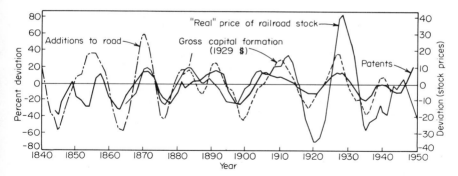

figure 4.6 *Railroad patents and railroad investment, United States, deviations of 7- or 9-year moving averages from 17-year moving averages. (Patents counted as in Fig. 4.4.)*

ented: chemicals, pharmaceuticals, electrical equipment, electronics, office equipment, communications, and aircraft have paced the United States economy (and the stock market) in the past two decades. These same industries have also often invested more of their own resources in research and development than the economy as a whole. Table 4.1 shows the relative research and development expenditures and innovativeness of major industries.

It is easy to assume that the economic success of very innovative firms like Xerox or IBM is a consequence of their devotion to fostering new technology. However, as Schmookler suggests, there are strong indications that the interaction is by no means a one-way street, and that success in the marketplace in turn stimulates further invention and innovation. Indeed, the mechanism of this interaction is clear: in a dynamic area where research is paying off, more research money and talent are quickly attracted to the field and, conversely, if research in an area is unproductive, it will just as quickly cease to attract investors.

MICROECONOMICS

At the level of competitive strategies of basic economic entities (e.g., corporations), it is worth noting that explicit managerial philosophy probably has as great an impact on the rate and direction of technological development as conversely. To take an extreme case,

TABLE 4.1

Industry	Research and development expenditures [9]		Innovation: % of total sales from new products [10]	
	% of net sales 1959	% of value added 1958	1961–1965	1966–1969 (planned)
Aerospace.................	20.8	30.9	8.5*	10.0
Electronics...............	12.8	22.4	} 5.5	6.0
Other electrical machinery..	10.1	16.3		
Instruments†.............	8.3	9.9		
Chemicals................	4.3	6.9‡	4.0	4.5
Machinery	4.2	6.3	5.8	5.8
Autos, trucks, and parts.....	3.4	10.2	2.5	5.5
Rubber	2.0	2.7	1.5	1.0
Metal products............	1.7	1.3	4.5	4.3
Store clay and glass........	1.4	1.2	3.3	4.3
Nonferrous metals.........	1.0	2.0	2.3	2.3
Paper and pulp...........	0.8	0.9	2.5	1.8
Ferrous metals............	0.6	0.8		
Textiles and apparel.......	0.5	0.2	3.3§	3.3§
Lumber and furniture......	0.5	0.2		
Food and beverages........	0.3	0.5	3.0	2.8

* Includes ships and railroads.
† Innovation figures included under metal products.
‡ Excludes petroleum refinement.
§ Textiles only.

a major company with an effective monopoly in its field of activity
may elect to suppress an important invention (if it can), rather
than adjust its manufacturing operations, marketing techniques,
etc. to the new technology.[7]   Although most of the popular stories
about allegedly suppressed inventions such as high-energy motor
fuels, long-lasting storage batteries, and long-lived light bulbs are
undoubtedly mythical, there can be no question that in some in-

---

[7] Alec Guinness' famous movie, *The Man in the White Suit,* was, of
course, a beautiful satire on this kind of situation.

dustries innovation is passively or even actively discouraged. Almost everyone is aware of the recent episode of the stainless steel razor blade, which was successfully introduced by a complete outsider to the razor blade industry (Wilkinson Sword Ltd.). There is little doubt that such a blade had been technologically feasible for at least a decade, if not more. A less well-known instance occurred not many years ago, when an outsider began marketing ammonium nitrate (fertilizer) as a cheap commercial explosive. This possibility had been known to insiders for at least 50 years, but had been quietly "overlooked" by the small, close-knit group of manufacturers who dominated the explosives industry. The present refusal of the automotive industry to recognize the impressive capabilities of modern external-combustion (steam) engines, is another case in point. Although steam engines would be quieter, cleaner, and cheaper than internal-combustion engines [11], they offer no advantage to the major manufacturers. Hence it is more convenient to go on pretending that steam engines are "obsolete."

The United States antitrust laws (and their active enforcement) are a major deterrent to suppressing inconvenient inventions by agreements "in restraint of trade." It is likely that one of the major advantages United States industry has had in competing against European companies with lower labor costs has been the European cartel tradition, which tended to produce management oriented toward maintaining the status quo by dividing existing markets, rather than looking for new sources of income. It might be noted here that the situation sometimes works in reverse. The German steel cartels were broken up after World War II by the Allies, and, since then, most of the coal and steel industries of Western Europe have been supervised by the joint Coal and Steel Community (CECA) which enforces a modicum of competition. On the other hand, the steel industry in the United States grew up under the protective umbrella of the U.S. Steel Corporation (which originally was assembled by J. P. Morgan as a monolith having 96 percent of the iron and steel business in the United States) and has never made any pretense of serious internal competition. The United States steel industry has never been very innovative and, as a consequence, has tended to adopt new methods only after they were invented and developed abroad. The most recent importation is the so-called "oxygen process" which was well established in Europe before being tried out to any great extent in this country. In 1953, when European industry was still rebuilding and reorganizing, United States export prices for steel averaged $104 per metric ton, compared to $115 for the members of the coal and steel community. By 1961 the ratios were reversed: $127 per ton for United States producers versus $99 per ton for the Europeans. Consequently the United States currently imports some 6 million tons of steel from Europe each year more than its exports—one of our most persistent balance-of-trade deficits.

Despite the above and a few other instances of industries still apparently dominated by a "competition-minimizing" philosophy, the majority of companies depend upon other approaches for survival. Three discernible attitudes worth distinguishing by a shorthand description are "cost minimizing," "sales maximizing," or maximum penetration of existing markets with existing products and emphasis on new products and/or new markets [12]. The third strategy tends to require a "performance-maximizing" approach. Most concerns have adopted combinations of these, but a few examples will suffice to illustrate the essential differences. The prime exponents of the cost-minimizing approach are the electrical utilities and the United States automotive industry, where the techniques of mass production were first applied and carried to their highest pitch of perfection. Until very recently, research and development was explicitly directed for the most part at reducing production costs, since even a few pennies saved on a part to be produced by the million can make a substantial difference to profits. Hence, long-range research has traditionally received very little support in comparison with the immense financial resources of these industries.

The auto industry today is also, to some extent, a sales-maximizing one.[8] Another exponent of the latter approach is the consumer products industry. Companies like Lever Brothers, Proctor and Gamble, Revlon, and others typically spend very large fractions

[8] More so than in the early 1920s, when Henry Ford refused to offer the Model T in any color but black (to keep costs and prices down).

of their total revenue (25 percent and up) on advertising and product development. Although they would like the customer to believe they have large, beautifully equipped research laboratories fully of chemists in white coats looking for better detergents, the truth is somewhat more pedestrian. Most of the research is devoted to testing consumer reactions to innovations in packaging, brand names, slogans, and sales appeals. What results is mainly new images (and minor variations) of old products. This competitive approach is actually not very conducive to rapid technological progress.

The innovative (performance-maximizing) industries are found mainly among those who sell to industry or to the government. Here the prime examples are aerospace, chemical, communications, electronics, and scientific instruments. Companies in this category are also among the fastest growing and most successful of recent decades: IBM, Control Data, Lockheed, Martin-Marietta, North American Aviation, Polaroid, High Voltage Engineering, Varian Associates, Texas Instruments, Fairchild Camera and Instrument, Xerox Corporation, General Telephone and Electronics, etc. These organizations rely primarily on advanced technology for competitive advantage and therefore concentrate major resources on fostering research and development.

Clearly a shift in the basic competitive strategy of an industry is likely to have a strong influence on the rate of technological progress in that industry. Strategies which were successful in the 1920s may not work out well today, and today's successful approaches may not be valid in the future. However, if research-oriented (performance-maximizing) strategies continue to pay off in terms of rapid growth and high profits, as they seem to have done in the recent past, more companies are likely to take this route. Of course, as was remarked earlier, the observed juxtaposition of rapid growth and rapid technological progress does not logically imply that the one is caused by the other; whence a company in a sluggish industry might pour money into research and achieve very little except disillusionment. This may have happened already in some instances and may be repeated again in the next decade or two. Nevertheless, it is clear that technological progress does, in some measure, depend on prevailing corporate competitive strategy, which is, in turn, conditioned in some degree by nontechnological factors.

COMMUNICATIONS AND SOCIAL FEEDBACK

As most people are well aware, society is undergoing an "information explosion." Not only has the quantity of information propagated each year grown very rapidly for the last 500 years—beginning with the printing press and followed by the telegraph, telephone, radio, television, communications satellite, and computer, as well as advances in personal mobility—but also there have been

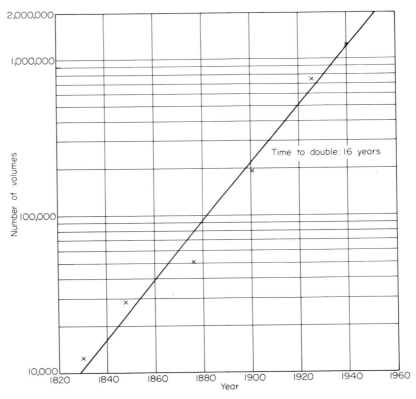

figure 4.7   *Average holdings of 10 libraries.*

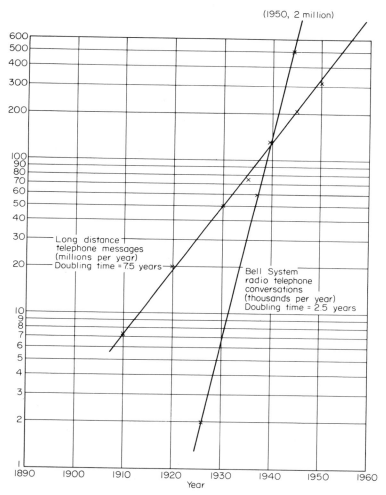

figure 4.8   *Growth of telephone communications.*

important resulting changes in the quality of societal interactions. Indeed the study of information flow has become an established branch of mathematical physics in its own right since the pioneering work of Shannon and Weaver in the late 1940s [13], while the social impact of the new forms of communication has been explored most deeply by Marshall McLuhan [14].

Quantitative indices of the growing quantity of information—

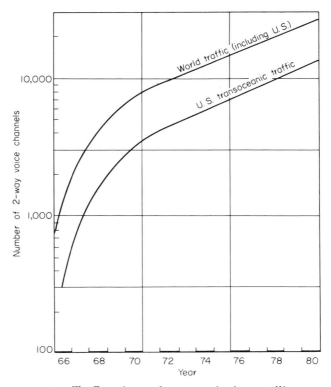

figure 4.9   *Traffic estimates for communications satellites.*

especially technological information—being generated each year are easy to find.   Thus in Fig. 4.7 are plotted the growth of average holdings in 10 university libraries in the United States [15]; in Fig. 4.8 we see some data on telephone messages [16].   Figure 4.9 shows some industry forecasts made in 1962 of the use of communications satellites [17].   One can find example after example to show that the rate of person-to-person information flow (both one-to-many and one-to-one) is increasing far faster than population, energy consumption, gross national product, or virtually any other major socioeconomic parameter.[9]   In some ways this is the the central fact of our times.

One is tempted to correlate the swelling flood of information

[9] Except possibly the destructiveness of weapons.

with the rapid growth of science as an institution in the last three to four centuries. Derek de Solla Price has documented this growth: thus Fig. 4.10 plots the number of scientists at any given time [18], while Fig. 4.11 shows the number of scientific journals published over the same period [19]. Scientists in later periods clearly have access in principle to more and more information; moreover, as the absolute quantity of raw information increases, there are ever greater incentives to filter it, digest it, sort it into categories, and boil it down more efficiently to make it more widely available. (Hence the growth of abstracting services and journals, for instance; automated *information retrieval* has also already become an important adjunct to bibliography.) While no scientist can read everything, even in his field, he does not have to rely on random selection: many factors combine to permit him to acquire relevant information more and more efficiently. Thus the social feedback process appears to be accelerating.

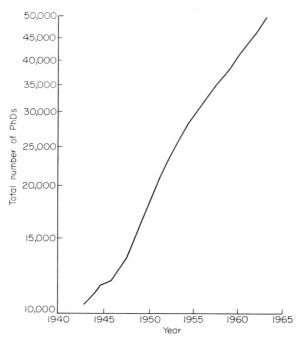

figure 4.10   *Number of Ph.D.s in engineering, mathematics, and physical sciences.*

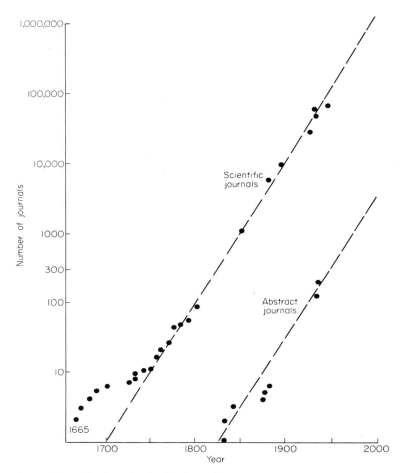

figure 4.11   *Number of scientific journals.*

TECHNOLOGICAL DIFFUSION
AND INNOVATION

The rate at which inventions are translated from the conceptual stage to important practical use is a measure of the operation of the feedback mechanism discussed above.   Gilfillan has investigated the lag time for a large number of important inventions. For instance, for a group of 19 "most useful" inventions introduced between 1888 and 1913 the time lag from "first thought" to first

working model or patent averaged 176 years.   The lag from thence
until first practical use averaged 24 years; commercial success required a further 14 years and "important use" another 12 years
[20].   Other groups of later inventions analyzed by Gilfillan
showed a distinct compression of the process, especially at the
earlier stage of evolution (what would now be called exploratory
development).   In recent years, especially, we are becoming accus-
tomed to a much more systematic and steadier progression, with
lag time related more or less directly to urgency and level of
investment.[10]

A recent study of technological diffusion by Frank Lynn, for
the National Commission on Technology, Automation and Economic Progress (1966), considered 20 significant innovations since
1880 [21].   The list is as follows:

Aluminum	Antibiotics
Motor vehicle transportation	Television broadcasting
Synthetic resins (plastics)	Titanium
Air transportation	Electronic computers
Vacuum tubes	Semiconductors
Radio broadcasting	Numerical control
Frozen foods	Nuclear power generation
Vitamins	Freeze-dried foods
Synthetic fibers	Integrated circuits
Synthetic rubber	Synthetic leather

Lynn defines the *incubation period* as the time between demonstrated technical feasibility and the recognition of commercial potential; the *commercial development* phase begins with the decision
to undertake such development and ends when the innovation is
introduced on a commercial basis.   For innovations reaching the
commercial stage prior to 1919 the average incubation period was
30 years, dropping to 16 years during the 1920–1944 period and
to 9 years since the end of World War II.   The total development
period dropped from 37 to 24 and finally to 14 years in the three
successive periods.

The rate at which a new technology penetrates a market is,

---

[10] Thus the very urgent "crash" program to develop the atomic bomb
required only 4 years to complete.

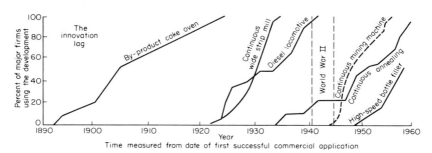

figure 4.12   *Market-penetration curve.*

of course, related to the social feedback described previously. There are two limiting factors: (1) the replacement cycle for existing equipment and (2) the learning curve[11] or acceptance curve characteristic of the industry or the market. The first is mainly determined by amortization and depreciation schedules (and, in some cases, tax policies). The learning curve, on the other hand, is a direct measure of the efficiency of communications and social feedback. At present we have no direct means of measuring the learning curve, in most cases, but it can be inferred from market-penetration curves such as shown in Fig. 4.12 [23] and a knowledge of the normal capital-replacement rates.

REFERENCES

1. Erich Jantsch, *Technological Forecasting in Perspective,* OECD, Paris, 1967.
2. Walter M. Miller, Jr., *A Canticle for Liebowitz,* J. B. Lippincott Company, Philadelphia, 1959.
3. *Physics Today,* October, 1962.
4. Louis Ridenour, *Bibliography in an Age of Science,* 2nd Annual Windsor Lecture, The University of Illinois Press, Urbana, 1951.
5. Raymond Isenson and Chalmers Sherwin, "Project Hindsight," First Interim Report, October, 1966.
6. *Scientific American,* September, 1963.
7. Jacob Schmookler, *Invention and Economic Growth,* Harvard University Press, Cambridge, Mass., 1966 and other references cited therein,

[11] This usage is unrelated to an hypothesis given the same name which was described by Jantsch [22]. The latter was related to the question of whether there exists a detectable systematic bias in the early stages of a serially updated technological forecast.

notably: "Economic Sources of Inventive Activity," *Journal of Economic History,* March, 1962.

8. *Ibid.*

9. Christopher Freeman, Raymond Poignant, and Ingvar Svennilson, *Science, Economic Growth and Government Policy,* OECD, Paris, 1963 (table 3, p. 80).

10. "Research & Development in American Industry," Department of Economics, McGraw-Hill publications, New York, May, 1966.

11. See R. U. Ayres and R. McKenna, "Technology and Urban Transportation: Environmental Quality Considerations," Resources for the Future and Hudson Institute, HI-949/1,2-RR, Jan. 11, 1968.

12. Robert A. Charpie, "Technological Innovation and Economic Growth," in *Applied Science and Technological Progress,* A Report to the Committee on Science and Astronautics, U.S. House of Representatives, National Academy of Sciences, 1967.

13. Claude Shannon, *Mathematical Theory of Communication,* The University of Illinois Press, Urbana, Ill., 1949.

14. See Marshall McLuhan, *Understanding Media: The Extensions of Man,* McGraw-Hill Book Company, New York, 1965 (and other publications).

15. Fremont Rider, *The Scholar and the Future of the Research Library,* Hadham Press, New York, 1944, cited by Ridenour, *op. cit.*

16. Ridenour, *op. cit.*

17. N. I. Korman, "The Communication Satellite: A High Power Satellite," Institute of the Aerospace Sciences, Paper No. 62-192, October, 1962.

18. *International Science and Technology,* September, 1963.

19. *International Science and Technology,* March, 1963.

20. S. C. Gilfillan, *Sociology of Invention,* Follett Publishing Company, Chicago, 1935; "The Prediction of Technical Change," *Review of Economics and Statistics,* vol. 34, pp. 368–385, 1952.

21. F. Lynn, "An Investigation of the Rate of Development and Diffusion of Technology in Our Modern Industrial Society," in "The Employment Impact of Technological Change," Appendix vol. II of *Technology and the American Economy,* The National Commission on Technology, Automation and Economic Progress, Washington, D.C., February, 1966.

22. Jantsch, *op. cit.*

23. *International Science and Technology,* December, 1963.

5 MORPHOLOGICAL
ANALYSIS

The descriptive label "morphological method" was coined by Fritz Zwicky, the famous astrophysicist and jet engine pioneer, to describe a technique for identifying, indexing, counting, and parametrizing the collection of all possible devices to achieve a specified functional capability [1]. Though not so used by its progenitor, the method can be used for identifying and counting all possible means to a given end at any level of abstraction or aggregation.

Perhaps the obvious application is in the analysis of technological opportunities: apart from the chance of using the scheme to anticipate actual inventions, there is at least a possibility of parametrically characterizing the optimum configuration for a particular mission or task. Whether or not this can be done with any confidence (as Zwicky asserts it can) one can at least devise a (partial) ordering of possible inventions in terms of their relative immediacy, other things being equal. This exercise is not a forecast per se, but it is a useful organizing tool, a source of insights, and a starting point for further analysis by other methods.

The basic method and its variations can best be explained in terms of specific illustrations. The simplest and perhaps "purest" example comes from Zwicky himself; he focused attention in great detail on the totality of all jet engines operating in a pure medium (vacuum, air, water, earth) containing simple elements only and being activated by chemical energy [2]. The following parametric possibilities were noted (the number of alternatives being given in parenthesis).

$p_1{}^{1,\,2}$	intrinsic or extrinsic chemically active mass	(2)
$p_2{}^{1,\,2}$	internal or external thrust generation	(2)
$p_3{}^{1,\,2,\,3}$	intrinsic, extrinsic, or zero thrust augmentation	(3)
$p_4{}^{1,\,2}$	internal or external thrust augmentation	(2)
$p_5{}^{1,\,2}$	positive or negative jet	(2)
$p_6{}^{1,\,\cdots,\,4}$	possible thermal cycles (adiabatic, isothermal, etc.)	(4)
$p_7{}^{1,\,\cdots,\,4}$	medium (vacuum, air, water, earth)	(4)
$p_8{}^{1,\,\cdots,\,4}$	motion (translatory, rotatory, oscillatory, none)	(4)
$p_9{}^{1,\,2,\,3}$	state of propellant (gas, liquid, solid)	(3)
$p_{10}{}^{1,\,2}$	continuous or intermittent operation	(2)
$p_{11}{}^{1,\,2}$	self-igniting or non-self-igniting propellant	(2)

There are 36,864 distinguishable combinations of the above,[1] although some are self-contradictory: thus the distinction between internal and external is meaningless for the case of zero thrust augmentation and extrinsic chemically active mass cannot apply if the medium is a vacuum. Allowing for all such restrictions, the number of possible jet engines appears to be 25,344. Obviously many of these permutations have never been tried out or

---

[1] This number is arrived at by multiplying $2 \times 2 \times 3 \times 2 \times 2 \times 4 \times 4 \times 4 \times 3 \times 2 \times 2 = 2^{12} \times 3^2$. If the $p_k{}^j$'s were written out in eleven rows, with two, three, or four columns respectively, a line connecting one element in each row would define one unique combination. The number of different such lines is exactly 36,864.

even thought about; i.e., they have not been "invented" yet.   Even if we disregard jets designed to operate in water or in the earth (hydrojets and terrajets), there remain more than 10,000 possible configurations of which only a small number have ever been given serious attention.   By picking out specific combinations—presumably more or less at random—Zwicky was able to suggest several rather radical new conceptual inventions such as the "aeroduct," a ramjet utilizing the chemical energy of free radicals and excited molecules in the earth's upper atmosphere, and the "aeropulse" or "rocket pulse," which carries part of its own oxidizer and obtains the rest from the outside air during the negative pressure phase of its cyclic (intermittent) operation.

The rules of morphological analysis, as originally expounded by Zwicky, are as follows [3]:

1. The problem to be solved or the functional capability desired must be stated with great precision.

2. The characteristic parameters must be identified.   To some extent this process is an automatic consequence of an exact definition of the problem, although this assurance may not offer much practical assistance to the analyst.   There are not, for instance, any reliable cookbook recipes for determining whether the list is *complete*.   (The difficulty of this may be pointed up by the observation that Zwicky's list is not complete: an important distinction can be made between the case where combustion occurs in a subsonic mass flow and the case where combustion occurs at supersonic speeds.   An embodiment of the latter concept is the supersonic combustion ramjet, or scramjet, one of the more promising configurations under consideration at the present time.   The addition of this dichotomous variable would roughly double the total number of possibilities enumerated above; if a further distinction were made between supersonic and hypersonic velocities, the total number of configurations would increase by a factor of 3.   Obviously, one can't be sure that this addition completes the list either.)

3. Each parameter must be subdivided into distinguishable cases or "states," $p_k{}^1$, $p_k{}^2$, $p_k{}^3$, . . . , as illustrated by the example of the jet engines quoted previously; or—more often—there is a continuum of values which must be classified into mean-

ingful ranges or regimes. Thus subsonic and supersonic speeds are distinguished by a well-defined discontinuous boundary (the speed of sound), but supersonic and hypersonic regimes—while different in important ways—are not sharply divisible one from the other.

4. Some "universal" method of analyzing the performance of the various combinations is needed. Zwicky concedes: "This is not always an easy task" [4]. It would be more realistic to say that it is seldom possible, in practice, if the subject area is at all complex. The jet engine utilizing chemical fuels is, after all, only one kind of jet engine: nuclear (fusion or fission) jets, ion jets, plasma jets, and even "photon jets" may easily be envisioned. If one goes further and asks for the totality of all kinds of devices converting energy from nuclear, chemical, or electromagnetic forms into kinetic energy of motion, the number of possible detailed configurations becomes astronomical—and we are still dealing merely with engines! The morphological map continues to encompass more and more territory as we include other energy-conversion and storage systems such as batteries, fuel cells, generators, refrigerators, and so on. Energy transmission opens up a whole new territory ranging from gears, belts, and axles to wires, waveguides, superconductors, and lasers.

If one attempts a schematic outline of the major possible ways of propelling a vehicle, for instance—even limiting oneself to broad classes such as "chemical-fueled jet," or "steam catapult" or "flywheel-electric"—one finds a surprising number of apparent combinations, as shown in Fig. 5.1 [5]. It is especially noteworthy that the boxes on each row are not each connected to all the boxes on the next row. Thus, there is no known way to charge a battery or to create a mass flow (jet) with a tightly wound spring. The lines which have been omitted are not all physically impossible combinations, however: for instance, it is perfectly feasible for nuclear combustion to support a photovoltaic cell. The connecting lines are omitted simply because the efficiency of such a converter would be ludicrously small.[2]

The type of morphological outline exhibited in Fig. 5.1 is primarily an elaborate checklist and device for organizing a broad

---

[2] Although this is similar to the mechanism of a solar cell.

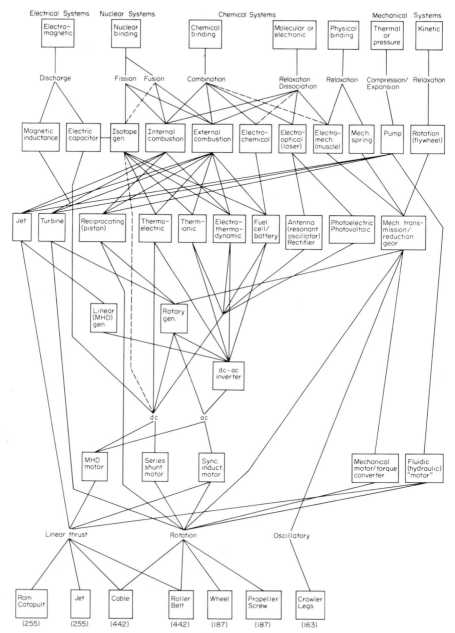

figure 5.1 *Sources of propulsive work.* (*Figures in parentheses below some items refer to the number of possibilities.*)

study [6]. However, morphological analysis can also sometimes be used to identify simple inventions which have hitherto been overlooked. This may seem surprising, in view of the very large number of inventors and inventions which have already been recorded. However, invention has hitherto been less than systematic: "hits" and "misses" are, evidently, to some extent random selections from a large collection of possibilities. The "screening" process which results in the acceptance and development of some inventions and the rejection of others, is not 100 percent efficient.

Figure 5.2 provides an illustration of the use of morphological techniques to identify opportunities which may have been previously overlooked [7]. In studying possible electric motor configurations one quickly notes that the rotor (and/or stator) may be passive and nonmagnetic, passive and permanently magnetized, or it may be an electromagnet. If the latter, it may be dc or ac, a monopole or multipolar. Altogether there are 6 possibilities for the rotor and 6 for the stator, or 36 in all. However, when these are examined individually, it can be verified that only 12

Rotor → / Stator ↓	Passive	Permanent magnet 2-pole	Electromagnet dc, 1-pole	Electromagnet dc, 2(N)-pole	Electromagnet ac, 1-pole	Electromagnet ac, 2(N)-pole
Passive	X	X	X	X	X	Inductive conjugate ?
Permanent magnet 2-pole	X	X	Homopolar (PM)	X	X	dc (PM) (with commutator)
Electromagnet dc, 1-pole	X	Homopolar conjugate (PM)	X	Homopolar conjugate (EM)	X	X
Electromagnet dc, 2(N)-pole	X	X	Homopolar (EM)	X	X	dc series/shunt (with commutator)
Electromagnet ac, 1-pole	X	X	X	X	X	ac homopolar conjugate ?
Electromagnet ac, 2(N)-pole	Inductive (squirrel cage)	Inductive synchro (PM)	X	Inductive synchro (EM) (slip rings)	ac Homopolar	X

figure 5.2 *Morphological summary of electrical torque-producing devices.*

of the combinations are physically capable of producing nonzero torque, as shown on the matrix.

One can easily identify many of the familiar types of electric motors, such as the ac inductive "squirrel cage" motor and the dc commutator motor(s). (It will be noted that the ac "synchronous" motor and the dc commutator motor with a permanently magnetized rotor are essentially conjugate versions of one another, with stator and rotor interchanged.) One of the most interesting entries on the table corresponds to an ac "homopolar" motor.[3] Although it is hard to prove that something is really "new," no previous reference to such a type has ever come to the attention of the author, at least. This possibility came to light only as a result of the use of a morphological approach [8].

The crucial difference between the analysis of a broad field and a narrow one is that many salient variables in the former case will not apply to all members of the class. To take an example already discussed, the class of all energy conversion devices encompasses both heat engines and electric motors. It makes sense to distinguish heat engines in terms of thermal cycle, continuous or intermittent operation, self-igniting or non-self-igniting, and so forth. On the other hand, electric motors would be distinguished in terms of some such typology as Fig. 5.2, broken down further into shunt-wound, series-wound, or compound categories. Neither set of variables makes any sense applied to the other type of system.

If a universal method of performance evaluation were really practical, one could pick out the optimum combination of elements for each designated application from theoretical considerations alone. Thus invention would be essentially replaced by straightforward analysis of alternate possibilities. Indeed, if the problem to be solved is sufficiently narrowly defined, there is actually some chance of "inventing" the solution on paper (as—presumably—in the case of the ac homopolar motor), although theoretical calculations are no substitute for actual realization.

The situation which generally prevails, of course, is that the

---

[3] This configuration might have at least one advantage over the more familiar dc version (or Faraday disk), namely the possibility of obtaining the necessary high currents and low voltages by means of a conventional transformer.

performance of untried combinations of elements is more or less uncertain, depending on the extent of departure from existing state of the art. It may be useful to think of the set of all possible combinations as a "lode" of interesting ideas, to use Holton's metaphor [9], or as a multidimensional "morphological map." The configurations which have already been concretely realized and are either in use in some form, or have been discarded, mark out an area of "occupied territory." Research and development is primarily devoted to the systematic and detailed investigation of the known territory on the "map," with the objective of improving upon the performance characteristics of existing devices. On the other hand, a small but significant fraction of the total research effort goes into exploration of the adjacent *"terra incognita."* It is the latter process which we wish to discuss now.

Given a definite starting point, namely a radical pioneering invention, there are several factors which combine to ensure that subsequent exploration usually tends to proceed from the known part of the morphological map only into the nearby territory. In other words, it is normal and natural to vary the parameters of the initial configuration one at a time, keeping the others constant. In this way a sequence of more or less favorable but similar arrangements is achieved.

Thus, historically, the open-cycle external-combustion reciprocating Rankine-cycle (steam) engine was followed by open-cycle external-combustion reciprocating Stirling-cycle and Ericsson-cycle engines using air as a working medium. Later these were followed by a steam turbine—later closed-cycle—on the one hand and by an internal-combustion reciprocating Otto-cycle engine on the other. The Diesel engine differed from its predecessor primarily in that it became self-igniting, while the obvious parallel of the external-combustion (steam) turbine was an internal-combustion version, the Brayton-cycle gas turbine. More recently still, the closed-cycle turbine has used other working fluids such as mercury vapor or heavy organics. Thus a small contiguous region on the morphological map of possible engines was exploited, as shown by Fig. 5.3. Yet all of these had in common the combustion of chemical fuels and the utilization of the resulting heat energy to cause a gas to expand and do physical work.

| | External combustion | | Internal combustion |
	Closed cycle	Open cycle	
Rotary	Closed-cycle vapor turbine (Rankine or Brayton cycle)	Steam turbine 'Simplex' rotary (Rankine cycle)	Gas turbine Free turbine Free piston- turbine (Brayton cycle) Wankel rotary Virmel rotary Tschudi rotary Mallory, etc. (Otto or Diesel cycle)
Reciprocating	Closed-cycle vapor or steam engine (Rankine cycle)	Steam engine: single acting or double acting (Rankine cycle)	Internal com- bustion engine 2-stroke 4-stroke
	Phillips engine (Stirling cycle)	Hot-air engine (Ericsson cycle)	(Otto or Diesel cycle)

figure 5.3 *Morphology of combustion engines.*

By the same token, a large number of alternative approaches to energy conversion were largely ignored, basically because they were *not* contiguous to the explored territory on the map. Thus, thermoelectric, thermomagnetic, thermionic, and electrochemical (galvanic) phenomena were, and are, largely unexploited as potential prime movers, especially in self-contained (i.e., automotive) applications.[4]

Accidents of history may have contributed in part to the order in which the various territories on the energy conversion map were explored, but it is no accident that exploration normally proceeds like an expanding inkblot from one morphological neighborhood to the next, rather than striking out "cross country," as it were. This is understandably rooted in the need to minimize both investment and risk in new enterprises, it being clear that both are strongly dependent on the magnitude of the changes required.

Let us now introduce some more precise definitions:

[4] Electric motors have been used, of course, for electric and diesel-electric locomotives, trolleys, subways, and some delivery trucks. However, electric engines for vehicles have never yet successfully competed with combustion engines for a large number of applications.

1. The *morphological space* of a broad area of technology consists of a set of discrete points or "coordinates," each corresponding to a particular combination of variables or parameters and each representable by a set of indices $\{p_k{}^j\}$. The space has as many dimensions as variables.

2. The *morphological distance* between two points in the space is the number of parameters wherein the two configurations differ from one another. Two configurations differing in only a single parameter are morphologically close together, while two configurations differing in many parameters are morphologically far apart.

3. A *morphological neighborhood* is a subset of points each of which is morphologically close to the other. Thus the subspace of internal-combustion engines constitutes a morphological neighborhood of the space of all energy conversion devices.

4. The *surface* of a morphological neighborhood is the set of all configurations differing in at most a single parameter from the points in the neighborhood. The *area* of the surface is the number of such points. A *weighted* area can also be defined by summing the numbers of points differing by one, two, three, etc., parameters, multiplied by appropriately decreasing coefficients, α_1, α_2, α_3, etc. An example illustrating the use of these weight factors will be given later.

5. Each time a new configuration becomes realizable in actuality, as a result of exploratory research and development, a *technological breakthrough* may be said to have been achieved. Thus a breakthrough is tantamount to developing new territory. On the other hand, refinements and improvements to a known configuration—however valuable—would not be characterized as breakthroughs in the above sense.

The probability of a breakthrough in a technological area, per unit time, is a decreasing function of its morphological distance from existing art, other things being equal. If the functional relationship is assumed to be an inverse quadratic, then a "distance" of 1 might correspond to a (normalized) a priori probability of 1, a distance of 2 would then correspond to a relative probability of $\frac{1}{4}$, a distance of 3 would correspond to a relative probability of $\frac{1}{9}$, etc. In any case, new developments will obviously tend to occur near older ones (in the morphological sense) essentially by accretion from the borders of state-of-the-art clusters into adjacent undeveloped regions.

Again, other things being equal, the opportunities available in a technological area at any given time will be approximately proportional to the surface of the (weighted) area corresponding to the surface of the occupied—or state-of-the-art—cluster. Assuming a quadratic inverse-functional relationship, as above, the weights would be equal to the a priori relative probabilities, namely, $\alpha_1 = 1$, $\alpha_2 = \frac{1}{4}$, $\alpha_3 = \frac{1}{9}$, and so on. The opportunities for technological progress (i.e. "breakthroughs") in the field will be scarcest at the beginning—when only a single point (configuration) is developed—and again toward the end when the entire morphological space has been thoroughly explored. In between there will be a point where the weighted surface of the occupied cluster reaches a maximum.

To make the foregoing discussion more concrete, we can postulate a morphological "space" which consists of the 27 combinations of values of the following three parameters:

$$\left. \begin{array}{l} p_1{}^j \\ p_2{}^j \\ p_3{}^j \end{array} \right\} j = 1, 2, 3$$

and suppose the configuration $(p_1{}^1, p_2{}^1, p_3{}^1)$ is the only one presently existing in the state of the art. Then the *surface* consists of the set

$$\{(p_1{}^2, p_2{}^1, p_3{}^1), (p_1{}^3, p_2{}^1, p_3{}^1), (p_1{}^1, p_2{}^2, p_3{}^1), (p_1{}^1, p_2{}^3, p_3{}^1),$$
$$(p_1{}^1, p_2{}^1, p_3{}^2), (p_1{}^1, p_2{}^1, p_3{}^3)\}$$

which comprises six points in all.

One can easily verify in a similar manner that there are 12 terms differing in two parameters and 8 terms differing in all three parameters. These are shown explicitly in the accompanying Table 5.1. For purposes of comparison, the table also shows the corresponding division when the occupied territory is a cluster of neighboring configurations. Clusters of 4, 6, 9 and 12 points are considered explicitly. It will be noted that the surface of the clus-

TABLE 5.1

Cluster of points occupied	Configurations differing in 1 parameter; points at distance 1 (perimeter)	Configurations differing in 2 parameters; points at distance 2	Configurations differing in 3 parameters; points at distance 3
(1) $(p_1^1 p_2^1 p_3^1)$	(6) $(p_1^2 p_2^1 p_3^1)(p_1^1 p_2^2 p_3^1)$ $(p_1^1 p_2^1 p_3^2)(p_1^3 p_2^1 p_3^1)$ $(p_1^1 p_2^3 p_3^1)(p_1^1 p_2^1 p_3^3)$	(12) $(p_1^2 p_2^2 p_3^1)(p_1^2 p_2^1 p_3^2)$ $(p_1^1 p_2^2 p_3^2)(p_1^3 p_2^2 p_3^1)$ $(p_1^3 p_2^1 p_3^2)(p_1^2 p_2^1 p_3^3)$ $(p_1^2 p_2^3 p_3^1)(p_1^1 p_2^2 p_3^3)$ $(p_1^1 p_2^3 p_3^3)(p_1^3 p_2^3 p_3^1)$ $(p_1^3 p_2^1 p_3^3)(p_1^1 p_2^3 p_3^3)$	(8) $(p_1^2 p_2^2 p_3^2)(p_1^3 p_2^2 p_3^2)$ $(p_1^2 p_2^3 p_3^2)(p_1^2 p_2^2 p_3^3)$ $(p_1^3 p_2^3 p_3^2)(p_1^3 p_2^2 p_3^3)$ $(p_1^2 p_2^3 p_3^3)(p_1^3 p_2^3 p_3^3)$
(4) $(p_1^1 p_2^1 p_3^1)$ $(p_1^2 p_2^1 p_3^1)$ $(p_1^1 p_2^2 p_3^1)$ $(p_1^2 p_2^2 p_3^1)$	(12) $(p_1^1 p_2^1 p_3^2)(p_1^2 p_2^1 p_3^2)$ $(p_1^1 p_2^2 p_3^2)(p_1^2 p_2^2 p_3^1)$ $(p_1^1 p_2^3 p_3^1)(p_1^1 p_2^1 p_3^3)$ $(p_1^3 p_2^2 p_3^1)(p_1^2 p_2^3 p_3^1)$ $(p_1^2 p_2^1 p_3^3)(p_1^1 p_2^2 p_3^3)$ $(p_1^3 p_2^3 p_3^2)(p_1^2 p_2^2 p_3^3)$	(9) $(p_1^3 p_2^3 p_3^1)(p_1^3 p_2^1 p_3^3)$ $(p_1^1 p_2^3 p_3^3)(p_1^3 p_2^1 p_3^2)$ $(p_1^1 p_2^3 p_3^2)(p_1^3 p_2^2 p_3^2)$ $(p_1^2 p_2^2 p_3^2)(p_1^3 p_2^2 p_3^3)$ $(p_1^2 p_2^3 p_3^3)$	(2) $(p_1^3 p_2^3 p_3^2)(p_1^3 p_2^3 p_3^3)$
(6) $(p_1^1 p_2^1 p_3^1)$ $(p_1^4 p_2^1 p_3^1)$ $(p_1^1 p_2^2 p_3^1)$ $(p_1^2 p_2^2 p_3^2)$ $(p_1^3 p_2^1 p_3^1)$ $(p_1^1 p_2^3 p_3^1)$	(15) $(p_1^1 p_2^1 p_3^2)(p_1^2 p_2^1 p_3^2)$ $(p_1^1 p_2^2 p_3^2)(p_1^3 p_2^2 p_3^1)$ $(p_1^2 p_2^3 p_3^1)(p_1^1 p_2^1 p_3^3)$ $(p_1^3 p_2^1 p_3^2)(p_1^1 p_2^3 p_3^2)$ $(p_1^2 p_2^1 p_3^3)(p_1^1 p_2^2 p_3^3)$ $(p_1^2 p_2^2 p_3^2)(p_1^1 p_2^2 p_3^3)$ $(p_1^3 p_2^3 p_3^1)(p_1^3 p_2^3 p_3^1)$ $(p_1^1 p_2^3 p_3^3)$	(6) $(p_1^3 p_2^2 p_3^3)(p_1^2 p_2^3 p_3^3)$ $(p_1^2 p_2^3 p_3^3)(p_1^3 p_2^3 p_3^3)$ $(p_1^3 p_2^2 p_3^3)(p_1^3 p_2^3 p_3^3)$	(0)
(9) $(p_1^1 p_2^1 p_3^1)$ $(p_1^2 p_2^1 p_3^1)$ $(p_1^1 p_2^2 p_3^1)$ $(p_1^2 p_2^2 p_3^1)$ $(p_1^3 p_2^1 p_3^1)$ $(p_1^1 p_2^3 p_3^1)$ $(p_1^3 p_2^2 p_3^1)$ $(p_1^2 p_2^3 p_3^1)$ $(p_1^3 p_2^3 p_3^1)$	(18) All remaining points		
. (12)	(15)		

ter reaches a maximum area when the cluster comprises from six to nine configurations, depending on the chosen weighting factor. The relation between cluster size and surface area—a measure of the available opportunities for further development—is shown in Table 5.2 for several possible a priori relative probabilities of a breakthrough as a function of increasing morphological distance. These probabilities form a series of decreasing weight factors, for nearest neighbors, next nearest neighbors, and so forth.

TABLE 5.2

Cluster size	Surface area weights: $\alpha_1 = 1$, $\alpha_2 = \alpha_3 = \cdots = 0$	Surface area weights: $\alpha_1 = 1$, $\alpha_2 = \frac{1}{4}$, $\alpha_3 = \frac{1}{9}$, ...	Surface area weights: $\alpha_1 = 1$, $\alpha_2 = \frac{1}{2}$, $\alpha_3 = \frac{1}{3}$, ...
1	6	9.89	14.67
4	12	14.36	17.33
6	15	16.5	18
9	18	18	18
12	15	15	15

The effective "surface area" of the cluster evidently increases to a maximum before beginning to decrease as the cluster gets longer and begins to fill the whole space. This behavior is shown graphically by the two-dimensional schematic diagrams in Fig. 5.4. Thus the number of opportunities for possible breakthroughs tends to reach a maximum at some intermediate stage in the development of a field. To the extent that technological progress depends upon the availability of promising avenues of research one would expect this, too, to reach a maximum followed by a "saturation" as the possibilities for further advance are exhausted.

Qualitatively one can see from Fig. 5.4 that the curve of progress in a field is likely to have a stretched-out S shape. A phenomenological model developed by A. L. Floyd based on concepts akin to the foregoing (discussed in Chap. 7) tends to confirm this surmise.

The point worth stressing in conclusion is that, with a little labor, it is possible to construct a morphological space for any well-defined technology, and to pick out—with a minimum of ambiguity—those configurations which are currently developed. In general, these will tend to be found in clusters, making the problem of estimating the surface area (or opportunity function) relatively straightforward. With a greater amount of labor, it would also be possible

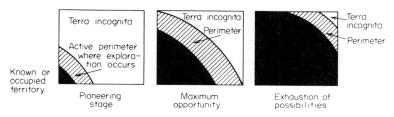

figure 5.4   *Progressive exhaustion of possibilities for invention in a field.*

to trace the history of the exploration and "occupation" of the territory in question. But, in either case, it should be possible to infer with reasonable confidence from the available information whether the point of maximum opportunity on the growth curve has yet been reached. As will be pointed out later, insofar as opportunity is related to growth, this is a rather crucial question in the sophisticated use of growth curves.

MORPHOLOGICAL ANALYSIS OF "FUTURE WORLDS"

In Chap. 8 some stress will be laid on the development of scenarios as an aid to intuitive thinking. Scenarios may be generated on the basis of combining historical events, more or less the way novelists "invent" plots; or they may arise naturally from role playing or gaming exercises.[5] It is also possible to generate scenarios systematically by morphological techniques. Variants of this method have been used quite explicitly at the Hudson Institute, and—in a disguised form—by many organizations (see Chap. 9). Clark Abt Associates have even gone so far as to incorporate computer "scenario generators" in certain of their large-scale sociopolitical-military simulations [10]. A very similar approach to mass pro-

---

[5] Each sequence of moves in a strategic or management game is, in effect, a scenario.

ducing scenarios has been taken by Theodore B. Taylor in a recent exploratory study of possible future nonnational nuclear threats [11].

For instance "conflict" scenarios may involve (almost) any combination of the following elements

$P_1^{1,2,\ldots,n_1}$ Regions: (Europe, Middle East, etc.)

$P_2^{1,2,\ldots,n_2}$ Level of conflict (subcrisis maneuvers, intense crisis, insurgency war, conventional war, nuclear war, etc.)

$P_3^{1,2,\ldots,n_3}$ Reason for United States involvement (treaty partner of one/both sides, traditional "friend" of one/both parties, ideological commitment to one side, United States military bases threatened, raw material supplies threatened, etc.)

$P_4^{1,2,\ldots,n_4}$ Strategic objective (engineer a coup d'état, stalemate, encourage negotiation to solve problems, "punish" one side, military victory)

$P_5^{1,2,\ldots,n_5}$ Type of intervention (diplomatic, take to United Nations, trade boycott, trade embargo, blockade, send "advisors," supply weapons, air/sea interdiction, ground forces, tactical nuclear weapons, etc.)

$P_6^{1,2,\ldots,n_6}$ Tactics (propaganda, infiltration, search and destroy enemy forces, clear and hold, etc.)

$P_7^{1,2,\ldots,n_7}$ Missions (communications, logistics, reconnaisance, etc.)

While not all missions (for example) are pertinent to all types of intervention—far from it—it is nevertheless clear that one can generate an enormous number of possible scenarios by combining and permuting the elements.

NETWORK METHODS

The basic concept of a multidimensional "hyperspace," which has been employed in this chapter up to now, has a rather limited communications value due to the inability of many people to think in such abstract terms. A metaphorical description of technological development as an expanding "inkblot" in such a hyperspace may not convey much to such a person.   Often the same basic factors can be expressed in a form which makes their interrelationships clearer.   Thus almost everyone immediately understands the conception behind a network such as Fig. 5.1 or Fig. 5.5, even though the connecting lines in the latter case represent rather intan-

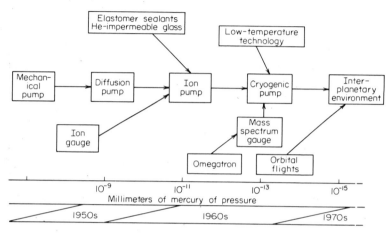

figure 5.5   *Trend as evolutionary development (example: high-vacuum technology).*

gible technological "inputs," which are neither strictly cause-effect nor are they strictly predecessor-successor relationships [12].   Nevertheless, Fig. 5.5 clearly conveys information about the evolution of high-vacuum technology as it did in fact occur.

Another type of network is illustrated in Fig. 5.6.   This procedure, commonly called *mission taxonomy,* is essentially a different but equivalent representation of the morphological breakdown of conflict scenarios described in the previous section [13–15].

Still another sort of network is shown in Fig. 5.7 [16].   This example is similar to Fig. 5.5, except that the direction of "flow" is reversed and the diagram focuses attention on outputs instead of inputs.   One can see at a glance how many and varied were the scientific offshoots of a particular set of experiments begun in the 1850s.

Working forward, instead of backward, in time, one can utilize either of these approaches.   For instance, if an objective is specified in the form of a desired capability, one can seek to identify:

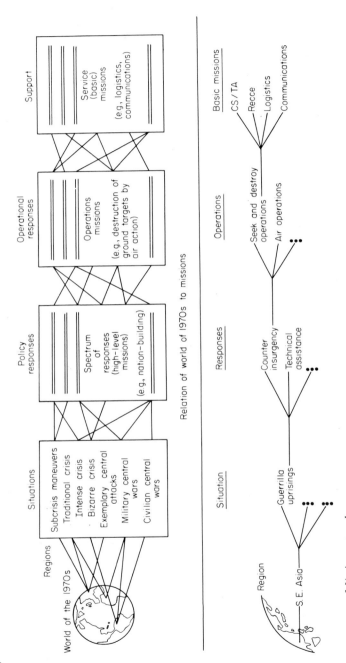

Relation of world of 1970s to missions

figure 5.6 *Mission network.*

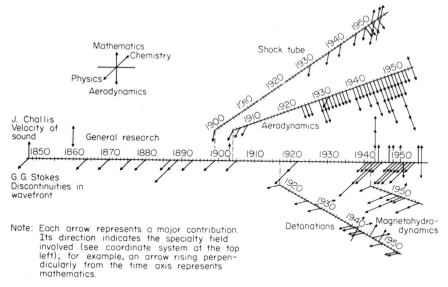

Note: Each arrow represents a major contribution. Its direction indicates the specialty field involved (see coordinate system at the top left); for example, an arrow rising perpendicularly from the time axis represents mathematics.

figure 5.7 *Possibilities arising out of research on wave propagation in compressible media: an historical research tree.*

1. Functional prerequisites (necessary conditions)
2. Alternative means of achieving them (sufficient conditions)

The systematic performance of step 2 is tantamount to a morphological analysis of the problem. If no specific goal is assumed, one can explore the various consequences which might follow from a particular hypothetical invention or development. In this case, the search would be for possible developments *beyond* the one in question, for which the latter is either a necessary or a sufficient condition. This activity would probably be little more than an intellectual game, however, unless the exploration itself is carried on for some definite purpose. One legitimate purpose might be to anticipate serious social problems which could follow on the heels of a particular discovery.[6] However, in practice, the end—or objective—is usually assumed and the investigation is focused on

---

[6] For instance, one can suppose that a breakthrough in catalysis might (?) lead to a method of manufacturing synthetic heroin. The example may be farfetched, but the principle is not. The discovery of a means of prolonging life indefinitely would lead to even more acute problems.

the various necessary conditions and alternative means of achieving it.

As an example, the Polaris submarine-based ICBM could not have been achieved without technological breakthroughs in six tangential areas, each of which was a functional prerequisite:

1. Nuclear reactors for marine propulsion
2. Accurate self-contained inertial navigation and guidance systems
3. Solid-fuel rocket boosters
4. Very compact thermonuclear warheads
5. Ablating reentry nose cones

The absence of any one prerequisite would have been enough to make the entire Polaris project infeasible.

Similarly, the electric automobile probably cannot be commercially successful without significant progress toward all of the following:

1. An improved electrochemical means of storing energy (battery or fuel cell)
2. Efficient solid-state electronic controls
3. Lightweight electric motors—probably ac
4. Effective means of recharging during a trip
5. A marketing system
6. A service and maintenance system

Morphological analysis suggests, of course, that several of the above requirements can be met by any of a number of alternative means or "equivalent inventions" in Gilfillan's terms [17]. For instance, recharging enroute could be achieved by *any* of the following:

1. High-rate secondary batteries capable of 15-min (or faster) recharge during stops (at specially equipped points on the trip)
2. Replaceable primary cells, or cells with replaceable anodes such as magnesium or lithium tape
3. Exchangeable secondary cells which can be physically removed and replaced by fully charged substitutes
4. On-board trickle chargers (e.g., a small two-cycle gasoline engine, or a fuel cell), either built in or rentable for trips

5. External power pickup (e.g., some inductive rf power supplies buried in the roadway)

In the case of a major departure from existing technology, with a very long gestation time, such as a high-speed underground transportation system, three-dimensional (holographic) point-to-point TV communications, controlled thermonuclear fusion, or interstellar space travel, one can sometimes identify longer branched chains of necessary and sufficient conditions. For instance, Fig. 5.8 illustrates part of a "contingency tree" for an underground high-speed transportation system.

Evidently a thorough analysis of the entire contingency tree of prerequisites and alternatives would be a fairly ambitious undertaking. It would, however, be very helpful in several ways. In the first place it provides a basis—if extended by the addition of organizational and other prerequisites—to assist in choosing the most promising fields for research support. For instance, high-power lasers, high-intensity pulsed electric and magnetic field research and plasma devices (which, in turn, requires high-power magnets) all appear at several points on the chart (Fig. 5.8). If one were to display the corresponding chain (not shown) of pre-

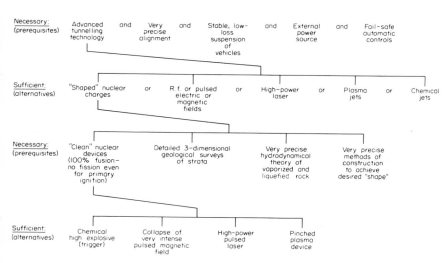

figure 5.8 *A contingency tree for a high-speed underground transportation system.*

requisites and alternatives with regard to suspension systems for vehicles in the tunnel, the possible use of high-power (superconducting) magnets would appear again. Similarly, lasers may be used for alignment purposes. Evidently these fields of research have a very high degree of relevance to the long-range objective of a high-speed underground transportation system—a fact which would not have been obvious to a casual observer.

The use of *relevance trees,* as they are often called, can be broadened to cover situations where the stated objective is much more general, such as antisubmarine defense or limited-war capability; or, instead of a single objective one may consider a number of objectives in parallel, with variable weighting factors. Questions which can be elucidated with the assistance of such a "tree" include (1) optimum allocation of fixed resources to maximize the utility of research and development for any given set of optimization criteria and for any definition of utility; (2) optimum selection of research and development program content, subject to the same conditions, or (in the case of industry) optimum choice of research and development proposals, from the standpoint of maximizing expected return—both in the form of revenue and technical cross-fertilization.

Where the objective is a specific one which has already been given official sanction and become a *program*—as in the case of controlled thermonuclear power—a tree of prerequisites and alternatives is the starting point for a critical-path network analysis (such as PERT) associated with a detailed plan of action. Here, of course, it is generally necessary to go into much greater detail, including not only intermediate research objectives but also decision points[7] and estimates of time, manpower, and budgets. These topics will be discussed again later in Chaps. 9 and 10, in connection with long-range planning activities.

REFERENCES

1. F. Zwicky, Miscellaneous papers (some unpublished, some published) written from 1938 on, but especially between 1943 and 1949 when Zwicky was director of research at Aerojet General Corp. A group

[7] This variant is sometimes known as a *decision tree.*

of these papers have been collected and reprinted under the title: *Morphology of Propulsive Power,* Society for Morphological Research, Pasadena, Calif., 1962.

2. F. Zwicky, "Tasks We Face," *Journal of the American Rocket Society,* vol. 84, pp. 3–20, 1951, reprinted in *Morphology of Propulsive Power, op. cit.*

3. *Ibid.*

4. *Ibid.*

5. R. U. Ayres and R. McKenna, "Technology and Urban Transportation: Environmental Quality Considerations" (2 vols.), Hudson Institute.

6. Apart from the study cited above, for instance, see Emile Hellund, "Fundamental Investigations of Electric Power Sources" (3 vols.), AFOSR-TR-59-103 (Defense Document Center, Accession no. 227,933).

7. R. U. Ayres and R. McKenna, *op. cit.*

8. *Ibid.*

9. Sara Dustin and Max Singer, "The Alternative National Policy Card Deck," vol. II of Edmund Stillman and Wm. Pfaff (eds.), "Study for Hypothetical Narratives for Use in Command and Control Systems Planning" (3 vols.), HI-285-RR, Hudson Institute, Inc., September, 1963. (Defense Document Center Accession no. AD 437,798.)

10. Robert Rea, "A Comprehensive System of Long Range Planning," in *Long Range Forecasting and Planning,* U.S. Air Force Office of Aerospace Research Symposium, August, 1966.

11. Theodore B. Taylor, W. R. Van Cleave, and E. M. Kinderman, "Preliminary Survey of Non-national Nuclear Threats," Stanford Research Institute Technical Note SSC-TN-5205-83, Sept. 17, 1967.

12. Thomas I. Monahan, "Current Approaches to Forecasting Methodology," in *Long Range Forecasting and Planning,* U.S. Air Force Office of Aerospace Research Symposium, August, 1966.

13. H. A. Linstone, "A Weapon Planning Problem for General Purpose Forces: A Functional Approach," RAND RM-3202, 1962.

14. H. A. Linstone, "MIRAGE 75," *Military Requirements Analysis Generation 1970–75,* Report No. LAC-592371, Lockheed Aircraft Corporation, January, 1965 (SECRET RD).

15. H. A. Linstone, "On MIRAGES," *Proceedings of the 1st Annual Technology and Management Conference,* J. Bright (ed.), Prentice-Hall Inc., Englewood Cliffs, N.J., 1968.

16. Gerald Holton, "Scientific Research and Scholarship: Notes toward the Design of Proper Scales," *Daedalus,* March, 1962.

17. S. C. Gilfillan, *Sociology of Invention,* Follett Publishing Company, Chicago, 1935.

6 EXTRAPOLATION
OF TRENDS

It is usual to let past experience be a guide for future expectations. The sun has always risen in the morning and set at night; ergo we expect it will do so tomorrow. The more often the inference has been tested and found to fit the observations, the more confidence we have in it. This increased confidence can, moreover, be expressed numerically, regardless of whether or not there is any theory to explain what is observed.

The above statement is straightforward enough when the record of past history forms a clear and unmistakeable pattern. Due to limitations of the human brain, or accumulated experience of a very broad kind, there are only two types of patterns which are instantly and universally recognizable, namely:

1. Periodic or cyclic "wave forms" with a single dominant period. Natural periodicities include the rotation of the earth (24

hours), the revolution of the moon around the earth (30 days), or the revolution of the earth around the sun (365 days), as well as the regular motion of a pendulum, or the beat of a pulse.

2. Continuous curves with at least two derivatives, at most one maximum or minimum and no points of inflection. The simplest examples would be a straight line, an exponential, a logarithmic curve, or any real function of the form

$$f(X) \propto X^p$$

where p may be positive or negative.

Any set of data points (i.e., a time series) which does not fit either of these descriptions may be a source of trouble. Many of the problems of extrapolation are essentially related to fitting unruly data into a mathematical straitjacket. The difficulty is as follows: given a finite number N of data points, there exists at least one polynomial (i.e., function of the form $A_0 + A_1X + A_2X^2 + \cdots + A_mX^m$) of degree $N - 1$ which will exactly fit the data, and—clearly—there is an infinite number of higher-degree polynomials, not to mention other continuous (transcendental) functions which also fit exactly. This infinitude of functions is not confined by any particular envelope, nor can one say anything particularly interesting about it as a class. In short, there is absolutely no reason to prefer any particular one of them to any other one.

The problem, clearly, is not to find a function which fits *exactly*—since there is no unique one anyway[1]—but to find a function which fits reasonably well and which is believable! This is not a requirement which can be stated rigorously; in fact, it can be seen that in the last analysis the choice will have to be made on the basis of nonmathematical (i.e., aesthetic) criteria. A rough description of what a curve fitter looks for in practice is: a continuous function which has very few maxima or minima and very few points of inflexion and which comes "close" to the majority of the data points in such a way that the positive and negative errors approximately cancel out. A more precise mathematical condition often used in curve fitting, *after* the functional form

---

[1] i.e., there are too many to choose from, all different.

of the curve has already been chosen, is to choose one of the disposable parameters by minimizing the *variance,* or the sum of the squares of the errors.[2] The choice of form is, however, generally not based on mathematical considerations at all: morphological simplicity, smoothness, and symmetry are probably the three fundamental guiding principles.

Since the above three aesthetic principles do not uniquely specify a curve, there is often room for an additional consideration of *mathematical* simplicity. This means that, other things being equal, we tend to try to fit empirical data to standard familiar functions such as exponentials, quadratics, or logarithms. In the case of periodic functions, of course, we tend to first try sines and cosines. In some cases, the presence of a recognizable element of statistical randomness in the data suggests the possible applicability of normal (gaussian) or log-normal functions.

At this point it seems pertinent to observe that there has been a proclivity among some forecasters and philosophers of forecasting to make a "gestalt" jump beyond the allowable arguments of simplicity, smoothness, and symmetry, to postulate a general "exponential law of social and/or technological progress." Henry Adams [1], Pitirim Sorokin [2], an Derek de Solla Price [3], among many others, have raised the exponential function virtually to the status of a Popperian "covering law."[3] Ridenour [4], Price, Gerald Holton [6], and others have also recognized that, in most cases, the exponential phase of growth eventually comes to an end, due presumably to "saturation" or the imposition of constraints. A convenient mathematical function which has this behavior is the so-called "logistic curve," of the form

$$f(x) \propto \frac{1}{1 + A \exp{(-kx)}}$$

The older literature of technological change is peppered with references to "logistic curves" or "S curves" (although there are

---

[2] Also called the "least squares" method.

[3] Every thesis also has its antithesis: in this case, there is also a school of thought which, as Ralph Lenz aptly remarks, "charges semi-logarithmic graph paper with possessing occult powers to distort honest data and extort false forecasts [5]."

many mathematical functions which have the saturation property), and the impression is sometimes left that the logistic curve has some more general significance. This is not the case, although a discussion of the S-curve phenomenology will be deferred.

Viewed more soberly, we may prefer exponential functions to fit time-series data (or gaussian functions to attempt to fit observed distribution functions), because there exists a convenient conceptual model, or mechanism, which has at least some superficial plausibility and from which the exponential (or gaussian) function can be deduced. Thus, an exponential growth function is an automatic consequence of a constant *rate* of change (i.e., birthrate or rate of invention).[4] One can adduce various arguments as to why it is reasonable to assume that the annual rate of increase should be constant, or approximately so. However, this type of reasoning is clearly the starting point for an iteration involving successive modifications of the assumed underlying mechanisms to fit the observed data, leading to ever more complex models—and deeper understanding—of the process of social change and technological progress. Projections based on such models have been classed as *heuristic* forecasts; this topic is reserved for the next chapter.

Apart from heuristic reasons for preferring exponential functions, there is another quite different reason for plotting data on semilog paper. Indeed, one of the (nonmythical) advantages of this technique is that it exhibits the deviations of the actual trend from an idealized exponential, of the form $f(x) \propto \exp(kx)$, which shows up as a straight line on a semilogarithmic scale. It is then very easy to discern whether the data has a systematic bias: whether the trend is "faster" or "slower" than an exponential, or whether the deviations merely tend to oscillate around a mean. If the trend is, in fact, an exponential, the relevant parameters can be determined directly from the intercepts.

---

[4] The most familiar example of all is the growth of a savings account at constant interest, compounded annually. If S_0 is the initial amount and r is the annual interest rate, the amount in the account after t years will be $S(t) = S_0 \exp(kt)$ where

$$k = \ln(1 + r) = r - \tfrac{1}{2}r^2 + \tfrac{1}{3}r^3 - \tfrac{1}{4}r^4 + \cdots$$

There are other standard transformations which may bring data into forms which provide useful insights. Although power laws of the form $f(x) \propto x^p$ are not at the moment much in vogue for representations of time-series data, they are often useful in describing empirical functional dependencies between variables (which might permit one time series to be derived from another, for instance). Be this as it may, any power law transforms to a straight line on log-log paper, and conversely a set of data which fits (or nearly fits) a straight line on log-log paper can be approximated by such a law.

Another function often used is the cumulative probability integral[5]

$$P(X) = \frac{1}{\sqrt{\pi}} \int_{-\infty}^{X} \exp(-u^2)\, du$$

which expresses the probability that a normal distribution will have a value less than or equal to X. If $P(X)$ is plotted on cartesian graph paper it turns out to be an S curve at least qualitatively similar to the logistic curve which has received such attention in studies of technological change. For statistical applications it is convenient to plot data on a transformed scale such that if the underlying distribution is normal (gaussian) the cumulative probability integral will be a straight line. Graph paper constructed in this way is called "probability paper." If the underlying distribution is log-normal,[6] a similar cumulative probability function can be defined which will look like an S curve on semilog paper, but which transforms to a straight line on "log-probability paper."

The use of probability or log-probability transformations does not logically imply any particular commitment on the part of the analyst to the basic assumptions of statistical probability theory—

---

[5] This is related to the more familiar error function as follows:

$$\text{erf } X = \frac{2}{\sqrt{\pi}} \int_0^X \exp(-u^2)\, du \qquad P(X) = \begin{cases} \frac{1}{2}(1 - \text{erf } X) & X < 0 \\ \frac{1}{2}(1 + \text{erf } X) & X > 0 \end{cases}$$

[6] That is, a normal distribution of the natural logarithms of the argument

$$N(X)\, dX = \frac{1}{\sqrt{2\pi}\, \sigma} \exp\left[\frac{-(\ln X - \mu)^2}{2\sigma^2}\right] \frac{dX}{X}$$

namely, that fluctuations around the mean (of the distribution) are due to independent, random events. If empirical data come very close to being a straight line on probability paper, the analyst might well suspect that a random process is involved somewhere in the background. But suspicion is not proof, and he need not interest himself in such questions at all. The elucidation of mechanisms may never arise if his only interest is straight extrapolation into the future.

However if an estimate of confidence level is required by the nature of the problem, then some statistical assumptions are needed: to wit, that errors, or deviations from the mean, are not "organized" or interdependent in any way and that no one of them depends on any other—in short, that they are randomly distributed. Given this assumption it is a straightforward matter to compute variances, standard deviations, and correlation coefficients. The basic data can also be manipulated in various ways to smooth out the effects of short-range fluctuations. Moving averages over various time periods (2, 3, 5, 10, or more years) often reveal cyclic perturbations with long periods—such as business cycles—which are not obvious to the average eye.

ENVELOPES, CONSTRAINTS AND SCALES

There are two common basic attitudes toward extrapolation in situations where the dynamism underlying a process of change is largely unknown but definite trends can be recognized in the time-series data:

1. One can assume a trend will not continue as before unless the known dynamics of the system are such as to clearly encourage its continuance. (System dynamics are rarely well understood, be it noted.)

2. One can assume a trend will continue as before unless the known constraints on the system clearly inhibit it.

The first attitude is unreasonable, if not irrational, because it insists that asymmetry between the future and the past is always to be expected, regardless of overwhelming evidence that the same assumption would almost always have been wrong at earlier times in history. On the other hand, the dangers of blind extrapolation have been summarized neatly by Ralph Lenz, to wit: it "errs by extending short-term trends, ignoring long-term trends, and forgetting common sense [7]." The importance of both constraints and common sense in forecasting may be illustrated by considering the probable result of a "naïve" extrapolation in 1967 of the trend of women's skirt lengths from 1957 to, say, 1975! The problem of forecasting constrained variables will be taken up later.

The risk of confusion of short-term and long-term trends can be minimized in two ways apart from exercising care in the analysis of constraints and external influences. Firstly it is necessary to extend the historical record backward in time as far and as accurately as possible. (The usual analyst's rule of thumb is that the projection forward should not exceed the time span of the baseline data.)

Secondly, as Gilfillan [8] and Lilley [9] have particularly emphasized, it is essential to avoid focusing too narrowly or attempting to predict specific details. Forecasters have repeatedly and unnecessarily made themselves appear foolish by succumbing to the seductive temptation to prophesy not only that some new capability will be achieved, but also *how*. For instance, H. G. Wells in *Anticipations* [10] foresaw the development of aerial warfare much as it has, in fact, developed—but strictly in terms of balloons! Increased mobility in land warfare was correctly foreseen, but only via the use of bicycles. Wells disparaged heavier-than-air craft, submarines, and tanks. In Chap. 2 this pitfall was described as overconcentration on known configurations, a particular sin of experts (especially engineers). Methodologically speaking, a better label for the fault might be *disaggregation*.

The tendency of disaggregative analysis is to consider each functional type of device (such as an airplane) parametrically in terms of its standard components: engine, propeller, guidance, airframe, fuel, etc. Progress for the performance of these subsystems may be considered separately. It is assumed that, together, these will put upper limits on overall performance.

Disaggregative analysis, however, is incapable of anticipating even the more ordinary kinds of innovation (if the system configuration is changed thereby), since the limiting performance of a given class of systems (devices) is derived by extrapolation only at the *component* level. Major inventions usually change the configuration of components in some way which cannot be foreseen in detail. Even abstract discussions which attempt to cover the universe of possibilities in terms of general categories, are prone to overlook important cases (which are often obvious enough in retrospect after one adjusts one's thinking).[7] In a sense, this omission is inevitable, since the only way to describe an invention in other than functional terms is to invent it first.

What, then, is an *aggregative* approach? In words, only functional characteristics of broad classes of devices should be projected. The ultimate definition of a broad class cannot, unfortunately, be other than relative: each class subsumes narrower ones, while it is included in broader ones; there is no absolute of narrowness or breadth, just as there is neither a smallest nor a largest positive number. The final decision as to whether or not a variable is suitable for projection can only be made in the context of the specific question being asked. These built-in ambiguities should not, however, obscure the essential differences between disaggregative, short-range, and narrowly focused projection on the one hand, and aggregative, longer-range, broadly focused projection on the other.

Figure 6.1, showing the historical development of particle accelerators for high-energy physics, illustrates the distinction between the aggregative and disaggregative procedures very well [11]. An analysis based on the design constraints on the configurations known at any given time would have predicted a maximum-energy curve whose rate of increase tapers off with time—similar to the

---

[7] For example, the intensity of light which can be concentrated into an "image" or spot was thought to be strictly limited by the optical theorem that the image cannot be brighter than the source. This theorem is correct, if rigorously stated, but it does not cover the possibility of amplification by stimulated emission (the laser).

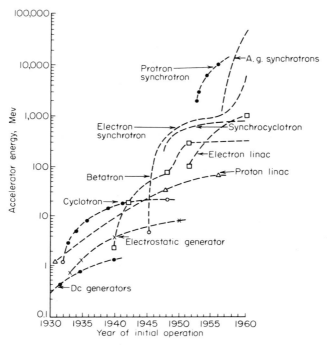

figure 6.1   *The rate of increase of operating energy in particle accelerators.*

actual curves for each type of machine.   Analysis of components and known configurations could not have predicted the *envelope* of the curves, however, since there was no basis for such a prediction; the essential information was lacking.   One could not, for example, have discussed the constraints on "strong focusing" systems such as the Alternating-Gradient Synchrotron (AGS) until somebody had formulated the essential idea.   The pattern of successive replacement of one class by another (called *escalation* by Holton [12]) is, of course, typical of an active field.   Figure 6.2 is another well-developed example of the same phenomenon [13]. In Fig. 6.3 it is even clearer [14].

Again, no amount of engineering expertise on reciprocating engines or gas turbines would have supplied an adequate basis for forecasting the envelope.   This is clearer if we consider a still larger class (*macro* class) of devices, such as was done in Fig. 2.2 for vehicle speeds. Nobody could discuss the engineering limits of flying machines before the Wright Brothers.

One can summarize the position simply by saying that the more disaggregative (component-oriented) the analysis, the more it is likely to be intrinsically biased toward conservatism. A forecast based on detailed knowledge seldom does what it purports to do, namely provide a firm upper limit to what can be achieved. In fact, it is almost normal for the maximum progress projected on the basis of forecasting the capabilities of components of a known configuration to be, in effect, the *lower* limit on actual progress, because although incremental improvements are assumed, it presumes no major inventions come along to change the basic way of achieving the desired result.

On the other hand, when we extrapolate envelope curves beyond the current state of the art, we automatically assume not only continued improvement and refinement, but also a continuation of the rate of invention which has characterized the socioeconomic system in the past. This will, of course, still fail to take into ac-

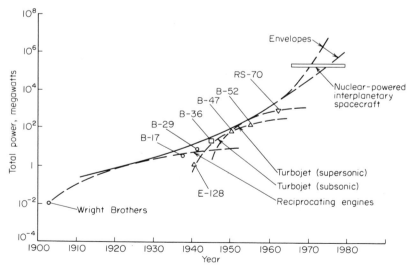

figure 6.2   *Aircraft power trend.*

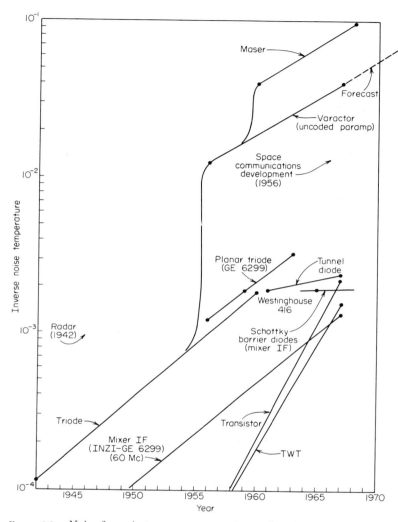

figure 6.3 *Noise figure improvement versus time and need.*

count the effects of very rare and extraordinary breakthroughs, but the consequences of a continuous process of "normal" innovation would presumably be taken into account.

The justification for this notion is, in brief, that the *past* performance of the (research and development) "system" is fairly likely to be a good model for its *future* performance; i.e., the system simulates itself.   One would expect this to be true, as long as, and to the extent that, there is no radical change in the environment.   For instance, consider the high-energy physics research establishment: it consists of individual scientists, institutional structures, and research tools and instruments having a considerable degree of continuity in time.   Thus, the physical equipment ages and is replaced at a relatively constant (or slowly increasing) rate; ditto the people.   In this light, it is plausible to expect the amount of creative work—which is equivalent to, or closely related to, the rate of invention—from year to year to increase (or decrease) relatively steadily and gradually, rather than fluctuating sharply up and down.   This characteristic rate of change is, of course, the slope of the envelope.

The case for the use of envelope curves rests so far on a heuristic understanding of what an envelope curve is.   It is useful to introduce some terminology and a few definitions.

A variable encompassing an aggregation of (more or less) independent component subvariables will be called a *macrovariable.* The subsidiary component variables may be called *microvariables.* This distinction is equivalent to the earlier relativistic dichotomy between broad and narrow.   Thus a microvariable with respect to a given macrovariable may itself be a macrovariable encompassing narrower microvariables.   In spite of this ambiguity, the usage is not likely to cause serious confusion in practice.

As mentioned previously, it is vital to pay careful attention to constraints.   The proper definition of constraint requires elucidation, however, since in a deeper sense all variables are constrained. One class of constraints is those intrinsic, natural (or physical) limits such as the speed of light, absolute zero temperature, absolute vacuum.   Ratios of output to input, measures of efficiency, or probability are also inherently limited to unity (or 100 percent). In addition, there are human limits imposed by the size, shape, and constitution of the human body, the nature of the environment and the earth.   Macrovariables whose rates of change are conditioned by or subject to any of the above limits will be called *intensive.*   A list of typical intensive variables is as follows:

Intensive Macrovariables

Energy conversion efficiency (e.g. electric generating efficiency)	
Information transmission through a specified channel	Ratios of input to output (or potential output) (units of energy, mass, volume, etc., not dollars)
Efficiency of harvesting crops	
Efficiency of recovery of metal from ore	
Efficiency of use of materials in fabrication, e.g., machining, stamping	
Pressures close to zero (high vacua)	
Temperature close to absolute zero	
Speeds close to the Einstein speed (velocity of light)	Absolute natural limits
Efficiency of a heat engine or refrigerator close to Carnot efficiency	
Maximum traffic flow per lane	Limitations due to human capabilities or characteristics of the earth (reaction time, toleration for g forces, noise toleration, gravity, circumference)
Maximum speed of elevator or escalator	
Minimum time between two points by plane	
Maximum velocity in atmosphere	
Minimum time to circumnavigate globe	

Another category of real constraints is absolute but time dependent natural limits imposed by existing capabilities in tangential areas, viz. other macrovariables. A macrovariable subject to a sliding constraint of this type will be classed as *extensive*. The following list provides some illustrations:

Extensive Macrovariables

Constrained parameter	Time-dependent constraint
Maximum dc magnetic field intensity	Strength of materials
Resolution of cameras	Film speed and lens focal length (f number)
Rankine-cycle efficiency	Strength of materials at high temperature and pressure
Laser coherence	Quality of crystals
Computer central processor speed	Miniaturization of electronic components
Refrigerator power and weight	Thermal insulation efficiency

In many cases, including some of the above, there is no single dominating constraint which, if removed, will result in a leap forward. In some situations a number of prerequisites (necessary conditions) are needed at once; in other instances forward progress can be achieved if any one of several limitations (sufficient conditions) are met. The interrelated *contingency trees* of necessary and sufficient conditions for progress (prerequisites on the one hand and alternative possibilities on the other) were discussed at the end of the previous chapter. It will be observed that, as a rule, a technological prerequisite for some advanced system would, in the present context, be described as removal of a time-dependent constraint. For example, it is reasonably clear that controlled fusion power cannot be practical without large superconducting (i.e., lossless) magnets and that present physical limitations on the size and performance of such equipment will have to be substantially exceeded.

The third important type of constraint is often unrecognized by noneconomists, but is no less real for that: it is the (time-dependent) limitation on performance imposed by the normal operation of a competitive marketplace. In simple terms: cost of production is, *ceteris paribus,* a function of level of performance. But production cost is in turn limited by marginal demand. Beyond a certain cost there will be no demand at all, and hence no production.

Let us consider the factors which determine the maximum speed of civilian airliners at any given moment. An obvious physical limit arises from the fact that frictional heating reduces the strength of structural materials, whence either the excess heat must be removed (for instance by ablation) or materials with greater strength/weight ratios at high temperatures are needed. But there is an absolute maximum strength/weight ratio for materials at any given time. (The best present-day materials are certainly not ultimate, but no one knows how much potential room for improvement remains.) But, in practice, the speed of airliners is not yet limited by this sort of problem. Performance of a newly designed airliner, such as the SST, would be determined by extrinsic economic factors, which are still more coercive at the moment, namely:

• Manufacturers' development costs, an increasing function of the degree of departure from existing technology

• Anticipated capital costs (per unit) to airlines, depending on size, complexity, departure from existing technology, and amortization over the number of units to be sold

• Fuel and maintenance cost as a function of size, speed, altitude, and length of average trip; also takeoff power and rate of climb (limited by ground-noise toleration)

• Income, depending on capacity, trip time, turnaround time, load factor (scheduling, traffic density, route structure, etc.); also government subsidies, if any

To take another example in a different field, the operating voltage of a long-distance electrical transmission line is not determined strictly by the optimum capabilities of conductors and insulators, but by:

• Development cost (an increasing function of the degree of departure from existing technology)

• Anticipated capital costs and amortization based on land acquisition (width of right-of-way depends on safety margin needed, hence on voltage), conductors, insulators, transformers, circuit breakers and possibly invertors (all voltage dependent)

• Anticipated operating costs and savings based on power, length of line, number of taps, and average load factor, etc. ("cost" are joule losses)

Obviously both time-independent and time-dependent constraints may exist, and one must decide which is dominant for the problem at hand. Thus, it is a useful exercise to classify macrovariables according to whether they are in extensive or intensive regimes.

The behavior of a trend curve as it begins to approach a constraint (i.e., in the intensive regime) is a major concern of the next chapter. However, it seems worthwhile anticipating the subsequent discussion to the extent of remarking that the transition between extensive and intensive regions often dictates a change of scale. Once a natural limit is near, further progress becomes progressively more difficult and units representing "equal effort" tend to become smaller and smaller.   In some cases, the relevant measure is a negative (instead of a positive) power of 10.

For example, progress in high-vacuum technology is typically measured in this fashion: a dozen years ago a vacuum of 10^{-10} torr was close to the limit on the state of the art, while currently one can achieve and measure 10^{-15} torr or better.   The latter figure is thought of as being five orders of magnitude better and would correspond to a shift of five decades on a logarithmic scale.

In low-temperature technology a similar situation prevails, although it is somewhat obscured by the standard choice of units (degrees on the Rankine or Kelvin scales).   In terms of average room temperature[8] the lowest temperature obtainable by means of a standard liquid helium cryostat is about 1.4×10^{-2} (or $4.2°K$), while the lowest temperature which can be maintained in a piece of commercially available laboratory equipment is 10^{-3} (or $0.3°K$).   On the other hand, the lowest temperature recorded, in a very elaborate and costly experiment, was 10^{-9} or one-billionth of room temperature, and one-millionth (10^{-6}) of the lowest level ordinarily achievable.   Again, in terms of effort, this difference is really six orders of magnitude—and would so appear on a logarithmic scale.

THE INERTIA OF TREND CURVES

An unevaluated assumption, which is somewhat basic to the systematic use of extrapolation techniques, is that macrovariables (more than microvariables) are likely to change relatively smoothly, continuously, and slowly without violent fluctuations.

One reason for this belief is that, almost by definition, a macrovariable involves a large number of interacting factors, including (but not limited to) individual decisions.   To the extent that these various factors and decisions are independent of one another, stability of macrovariables is partly a result of the operation of the "law of large numbers." This law, which is a weak form of the

[8] Conventionally assumed to be $300°K$ or $27°C$ (equivalent to $80°F$).

fundamental central-limit theorem of statistics,[9] says that an additive macrovariable (i.e., sum over many components) is likely to deviate relatively (percentagewise) less from its mean or *centroid* than any of its component microvariables, provided that the latter are all roughly equal in size and vary independently of one another within similar ranges.

It is tempting to use this theorem immediately in a model to describe the activities of aggregations of individuals. Thus it might be argued that fluctuations in gross national income (GNI) per capita will surely be smaller than the mean fluctuations of individual incomes. Because of many interactions (including the feedback mechanisms to be discussed later), the latter are *not* all independent of one another, which ruins the mathematical rigor of the conclusion, even though much of the "smoothing" still seems to be due to mutual compensations of many random fluctuations. To the extent that individual microvariable fluctuations are not independent, they must be correlated with each other and hence with changes of the macrovariable. Thus there will be two sets of fluctuations—individual and aggregate. Unfortunately, there is no fundamental reason why aggregate fluctuations need be small—we have examples in history, such as the depression, where they were not.

Nevertheless, it is a matter of everyday experience that macrovariables do normally fluctuate more slowly than microvariables. It is easy to convince oneself that the rate at which aggregate variables fluctuate is somehow a function of the reaction time (or "relaxation time") of the interconnected system to a perturbation. This determines the speed with which a disturbance (analogous to a ripple in water) will be propagated throughout the system. For example, when a company in a competitive industry (not domi-

---

[9] Suppose an additive macrovariable is defined as $P = \sum_i x_i$. Then the standard deviation of P is given by

$$\sigma_p{}^2 = \sum_i \sigma_i{}^2 \qquad \sigma_p = \sqrt{\sum_i \sigma_i{}^2}$$

If N of the σ_i are all of roughly equal magnitude (the others being negligible), and $\bar{x}$ is the mean value of the nonnegligible variables, then $P \cong N\bar{x}$ and $\sigma_p/P \cong \bar{\sigma}/\sqrt{N\bar{x}}$.

nated by a single firm) changes its price structure, the first reaction of its rivals is usually to sit back and wait to see what effect this move has on the pattern (i.e., the supply-demand relationship) of the market. This will take more or less time depending on a variety of circumstances. If the response seems to be favorable to the innovator, another company will take the plunge and sooner or later the rest will be likely to follow. If not, the company that started the move may return to the *status quo ante*. If the competing companies in the above illustration must change their technology to keep up with the innovator, their response may be even slower.[10] But this means also that when technical changes are reflected in economic terms the effects are filtered and spread out considerably in time—an important point. It is a fact that the practical use of technology often lags far behind laboratory demonstrations of feasibility.

As previously indicated, certain types of fluctuations are strongly correlated and therefore do not meet the statistical requirement of independence. There are many feedback mechanisms in operation, some of which tend to amplify, and some of which tend to smooth over disturbances. A typical amplification mechanism helped to produce the great stock market crash of 1929: many investors had bought on margin, using their stocks as collateral for loans. When stocks fell due to a wave of selling, so did the value of the collateral backing the loans; brokers or lenders were forced to cover loans by additional selling which, in turn, forced stock prices down further, etc. The converse of this amplification mechanism in the stock market is short covering. When many investors sell short (i.e., borrow stock and sell it, promising to repay the stock at a later date), and a stock rises anyway, the shorts are forced to buy heavily on a rising market—thereby pushing it up further.[11]

[10] For example, they may not elect to operate at a loss due to internal or institutional constraints such as the risk of loss of good credit ratings, or stockholder dissatisfaction, but to rely on customer loyalty and inertia to maintain sales volume near the old level during the changeover period.

[11] This is not merely a theoretical description of something that could happen: for instance the sudden boom in Comsat Corp. stock in December 1964 was due to panic covering by people who went short on or shortly after the original issue (thinking it was overpriced).

One can find analogous mechanisms in operation elsewhere in the economy. For example, a crisis of confidence (for any reason) can cause businessmen to cut back on investment, which in turn reduces the amount of money in circulation and the volume of business in certain industries, etc. A sufficiently convincing prophecy of economic downturn (or upturn) can, literally, be self-fulfilling.[12] One of the great lessons of the thirties was the need for creating (and using) additional damping (as opposed to amplifying) feedback mechanisms which can be put into operation to counteract recessions and keep them from becoming depressions, or—conversely—to cool off inflations. There are a great many such measures currently at the disposal of the government, including monetary policy (interest rates and money supply) and fiscal policy (tax rates and spending).

In the realm of technology, the purely economic feedback mechanisms—both amplifying and damping—are not very important. There are, however, analogous ones. The well-known "bandwagon effect" is an example of amplification. A discovery with dramatic impact, such as the laser, can lead to a sudden and excessive diversion of effort from other fields of investigation into the new one. For instance, by 1962, barely 2 years after Theodore Maiman's first demonstration of laser action in ruby, there were 350 organizations involved in laser development in the United States. This naturally produced a tremendous burst of progress in the new area, as shown in Fig. 6.4, which plots a five order of magnitude increase of output energies in 4 years [15]. There is also another side of the coin: when an applied research program meets an unexpected check it may be abandoned just as suddenly. Thus a tremendous wave of interest in superconducting thin films for high-speed computer logic and memory elements was stimulated as a result of Dudley Buck's invention of the "cryotron" at M.I.T. in 1956. By 1960 virtually every computer manufacturer and electronics manufacturer had initiated a project in this area.[13] Yet after 1962 IBM, Univac, and most of the other early optimists

---

[12] President Hoover was blamed bitterly for talking prosperity in a time of depression. This was to some extent unjust, for he was doing his best to create business confidence—unfortunately at the cost of public confidence.

[13] Notably "Project Lightning" at IBM, supported by the Defense Department.

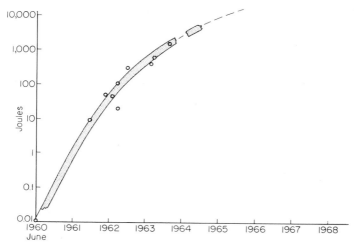

figure 6.4   *Single pulse energy, lasers.*

cut back their efforts or dropped out of the race—virtually leaving
the field to RCA—largely due to unexpected difficulties in fabricat-
ing reliable superconducting thin film components [16].

There are, similarly, some very potent factors which tend to
damp out the fluctuations of technological macrovariables, mainly
arising out of the exigencies of technological-economic competition.
Consider a situation where several research groups are actively
pushing the state of the art of some class of devices, such as radars
or computers. There are several design approaches and numerous
possible choices of components for each purpose. Each research
and development group keeps an eye on what the others are doing
(even though a considerable degree of secrecy may be imposed).
At some point one group decides it is ready to build a prototype
of its next generation entry. The question arises: what performance standard should the model be designed to meet? In practice, probably the safest way of determining this is to extrapolate
the recent rates of progress of the relevant state-of-the-art parameters
forward to the estimated time of completion of the prototype.

The reason for preferring this choice is that it appears to be
the optimum compromise between (1) a strong motivation to keep
up with the advancing state of the art and the competition and
(2) a strong motivation not to jeopardize the chances of success

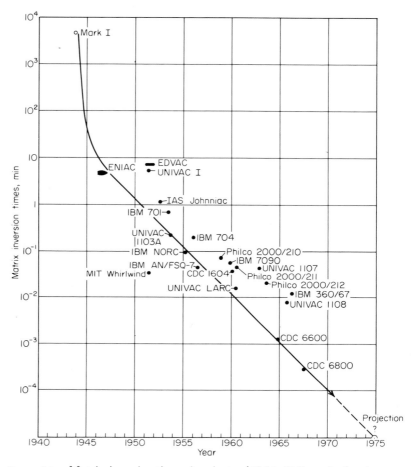

figure 6.5   *Matrix inversion times, in minutes (40 × 40 "standard estimate").*

by attempting to achieve too much too soon.[14]   Evidently here is an example of a self-fulfilling prophecy. (Furthermore, the more frequently this kind of criterion is used successfully in planning, the more confidence planners tend to have in it and the more reliable it will become as a guide.)

Hence if the envelope curves for macrovariables representing the state of the art of a given technology show a definite trend,

[14] This obviously resembles a "mini-max" or "saddlepoint" type of solution in the theory of games.

there already exists an internal dynamics which tends to continue the trend more or less straight ahead (on the appropriate scale) until some constraint is reached. Thus if the curve is rising exponentially at a given rate, it is likely—other things being equal, of course—to go on doing so. By the same token, if there has been little or no recent progress in the state of the art, there is less motivation for any competitor to force a change.

Figures 6.5 and 6.6 show the progress in an industry where forward design has almost certainly been strongly influenced by

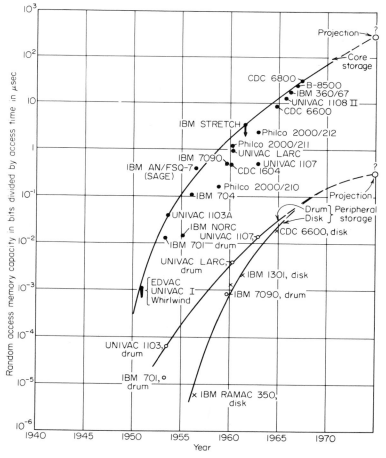

figure 6.6 *Random access memory capacity (bits) divided by access time* (μsec).

the trend of past performance [17]. It is noteworthy that the rapid progression in performance capabilities has continued through several radical changes in the underlying technology. It will be recalled that the Harvard Mark I (1944) was an electro-mechanical system without a stored program capability. The Eniac, Edvac, and Univac I utilized mercury acoustic delay lines for high-speed memory and vacuum tubes for logic circuits. The IBM 701, IAS Johnniac, and M.I.T Whirlwind utilized electrostatic memory storage devices (cathode ray tubes), which were rapidly supplanted in the Whirlwind and 701 by ferrite magnetic cores (still in use today). Transistors took over most logical functions by the mid fifties. Thin-film "scratch pad" memories were intro-duced by the Univac Larc in 1960. Hybrid and fully integrated circuits were introduced after 1965 by IBM and RCA respectively. The next generation of computers will probably have totally new capabilities, such as time sharing, which are not reflected at all in the parameters illustrated.

REFERENCES

1. Henry Adams, *The Education of Henry Adams,* reprinted by Heritage Press, New York, 1958.
2. Pitirim Sorokin, *Social and Cultural Dynamics,* Harvard University Press, Cambridge, Mass.
3. D. J. de Solla Price, *Little Science, Big Science,* Columbia University Press, New York, 1966.
4. Louis Ridenour, "Bibliography in an Age of Science," *2nd Annual Windsor Lectures,* The University of Illinois Press, Urbana, Ill., 1951.
5. Ralph C. Lenz, Jr., "Forecasting Exploding Technologies by Trend Extrapolation," in J. Bright (ed.), *1st Annual Technology and Manage-ment Conference,* Prentice-Hall, Inc., Englewood Cliffs, N.J., 1968.
6. Gerald Holton, "Scientific Research and Scholarship: Notes toward the Design of Proper Scales," *Daedalus,* 1962.
7. Lenz, *op. cit.*
8. S. C. Gilfillan, "The Prediction of Inventions," in W. Ogburn (ed.), *Technological Trends and National Policy,* U.S. National Research Council, 1937.
9. S. Lilley, "Can Prediction Become a Science?" *Discovery,* November, 1946. Reprinted in Bernard Barber and Walter Hirsch (eds.), *The Sociology of Science,* The Free Press of Glencoe, New York, 1962.
10. H. G. Wells, *Anticipations,* Harper & Brothers, New York, 1902.
11. M. Stanley Livingston, reprinted in Holton, *op. cit.*
12. Holton, *op. cit.*

13. D. G. Samaras, "Nuclear Space Propulsion: A Historic Necessity," *Nuclear Energy*, p. 352, September, 1962.
14. Karle D. Packard, Airborne Instrument Laboratories, Inc. (Division of Cutler-Hammer Corp.), personal communication, 1968.
15. See James R. Carter, "Lasers and How They Grew," U.S. Naval Ordnance Test Station: IDP-2109, August 11, 1967. Also: R. U. Ayres, "Significance of Laser Weapon-systems to the Arms Control and Disarmament Agency in the 70's." Annex 7 to D. G. Brennan (ed.), "Future Technology and Arms Control," HI-504-RR/A7, Hudson Institute, New York, June 1, 1965.
16. R. U. Ayres et al., *Hypercryogenics*, Science for Business, Inc. Published by Plenum Press, New York, 1965.
17. Data compiled by the author from a variety of sources and presented at the *2d Annual Technology and Management Conference*, Washington, D.C., March, 1968.

7 HEURISTIC
FORECASTS

On strictly empirical evidence, without understanding underlying causal relationships, it may occasionally be observed that movements in one trend curve closely follow movements in others. If this is so, the latter may provide a guide for (short-term) forecasts of the former. This method is used by some macroeconomists who forecast movement in the gross national product by looking at trends in housing starts, rail car loadings, automobile dealer inventories, steel production, and so forth. It is also used by certain stock market analysts who keep careful track of such indicators as odd-lot purchases and sales (the odd-lotter is assumed to be always wrong), cumulative short interest, ratios of bond interest to dividends, and so forth.

However, in technological forecasting, empirical relationships

118

such as the above play little role, if for no other reason than because detailed data are unavailable. There are, however, certain "structural" relationships between trends which can sometimes be used to assist in the refinement of projections. The logic is, essentially, that a good way of forecasting $X(t)$ may be to identify a functional relationship between $X(t)$ and some other variable $Y(t)$ which for some reason is easier to forecast. A special case, of course, is where $X(t)$ is simply proportional to $Y(t - T)$, that is Y is a *precursor* of X. In practice, unfortunately, this is not a common situation. The one obvious example, cited by Lenz [1], is the historical parallelism between military and civilian transport aircraft, illustrated by Fig. 7.1. This follow-the-leader relationship reflects cross fertilization and spin-off in the aircraft industry, where military research and development normally finds civilian applications after a few years. One might, at first sight, expect to find

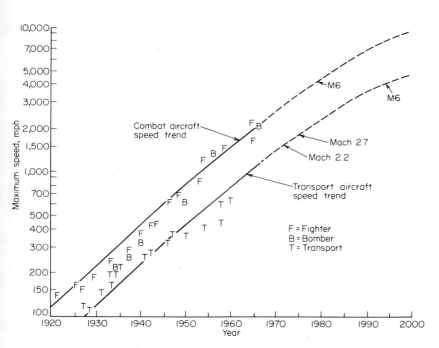

figure 7.1   *Speed trends of combat aircraft versus speed trends of transport aircraft, showing lead trend effect.*

other similar parallelisms between military and civilian technology, but if they exist they are elusive.[1]

There is another type of trend correlation, also discussed by Lenz, based not on technological transference, but on direct algebraic relations among variables. The fundamental notion is logically trivial: if $Z = YX$ then one can forecast Z by forecasting X and Y separately. Practical applications of this sometimes lead to unexpected insights, however. Lenz notes that domestic trunk airline passenger miles, plane miles, seating capacity, and load factor can all be extrapolated separately from time-series data, but that they are not all independent [2]. In fact

$$\text{Plane-miles} = \frac{\text{passenger-miles}}{\text{seating capacity} \times \text{load factor}}$$

Assuming a constant (60 percent) load factor and a continuation

[1] This is not intended to imply that technological spin-off is particularly rare, but only that military hardware seldom has direct civilian counterparts, except at the *component* level.

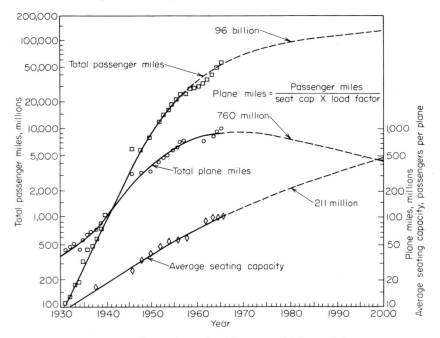

figure 7.2 *Domestic trunk airlines—multiple trend forecast.*

of the historical rates of increase of passenger-miles and seating capacity, one arrives at the somewhat surprising conclusion that plane-miles (hence the number of planes in the air) will reach a maximum in the early 1970s, followed by a decline, as shown in Fig. 7.2. The essential feature of this example is the correct identification of plane-miles as a dependent rather than an independent variable.

PROJECTION OF CONSTRAINED VARIABLES

Although constraints appear in a variety of forms and shapes, one type has already been encountered, namely the case where a variable is subject to an inequality, i.e., limited by a "floor" or "ceiling" (which may or may not change as a function of time). Examples of such variables were cited in the previous chapter. The characteristic saturation behavior of a trend curve as it begins to "feel" the approach of the looming constraint has apparently suggested several mathematical models which will be described in due course. However, these models generally have a broader justification and deserve separate discussion.

Projections of household expenditures under a budgetary constraint would be a simple case in point. Total income is the only independent variable, and allocations among various options, e.g., food, shelter, clothing, recreation, transportation, and education can only be predicted by means of a rather detailed knowledge of their indifference curves relative to each other.

It is noteworthy that this situation is totally different from the example cited previously where one of a group of four variables was dependent on the other three. The distinction is that, with regard to the budgetary limit—or any other similar situation where a condition exists on a functional of all the variables—*all* of the variables except one (total income) are interdependent on each other, whence *only* that one can be extrapolated safely.

Mathematically, any constraint of the form

$$f(X_1, \ldots , X_m) = 0$$

will have a similar effect, namely that independent treatment of the $X_1, \ldots, X_m$ is an illegitimate procedure; moreover, one cannot validly assume that one of the variables is strictly dependent on the others unless this assumption is justified by the actual relationships. An alternative approach to forecasting is needed, which may require much more detailed data and a correspondingly greater degree of understanding (i.e., at least a phenomenological model).

It is almost superfluous to cite examples of this type of behavior: indeed, interdependency among variables is the rule, independence the exception. The latter is often assumed for convenience and allowed by convention in many simplistic short-range forecasts, but the assumption must clearly be jettisoned whenever circumstances, as dictated by the specific problem, require it.

ANALOGIES, METAPHORS, AND STRUCTURAL MODELS

A good starting point seems to be Henry Adams, who conceived several familiar metaphors to describe technological progress [3]. At one point he likens it to the "formula of chemical explosion," and at another point he draws an analogy between the attractive

TABLE 7.1

Biological growth	Technical improvement
Initial cell	Initial idea or invention
Cell division	Inventive process
Second-generation cell	New idea or invention
Cell division period	Time required for initial invention to stimulate new invention
Nutrient media	Economic support for inventive process
Cell lifetime	Useful life of invention
Cell death, normal	Obsolescence of invention
Cell mass	Technical area or machine class
Volume limit of cell mass	Limits of economic demand for invention in a given technical area
Size of cell mass	Total of existing nonobsolescent inventions in technical area
Strength of cell mass	Performance capability

force of gravity and the "force" which draws "thought" (i.e., conception) closer and closer to nature (i.e., reality).[2] Adams essentially attributed the takeoff point of the modern age of science to a revolution in metaphysics: the fourteenth- and fifteenth-century rejection of the old notion of a priori knowledge—"evolving the universe from a thought"—in favor of a posteriori knowledge of an objective reality. Several times he remarks that science is doubling its complexities every 10 years, or words to that effect, which is equivalent to an exponential model of technological progress.

Ralph Lenz suggests that the use of such phrases as "father of the idea" or "fertile imagination" and words such as "embryonic," "growth," or "maturity" is evidence of widespread intuitive understanding of a fairly elaborate biological analogy. Lenz notes the comparisons in Table 7.1 [5].

Since the biological model, here, is a population of organisms, Lenz proposes the use of Raymond Pearl's formula to describe population growth as a function of time in a limited environment [6,7].[3] The Pearl equation applies equally well to fruit flies in

[2] This actually implies a quadratic rate of change, as Jantsch shows [4].

[3] There are two other well-known "growth" laws: the so-called Gompertz equation and the von Bertalanffy equation. They appear very different mathematically, although if the three curves are superimposed with equal slopes at the point of inflection, they are difficult to distinguish. Qualitatively, they all yield the characteristic S shape. A detailed comparison is as follows [8]:

	Logistic (Pearl)	Gompertz	von Bertalanffy
Form, P	$\dfrac{1}{1 + A \exp(-kt)}$	$\exp[-Ak^{-kt}]$	$[1 - A \exp(-kt)]^3$
Slope, $\dfrac{dP}{dt}$	$KP(1 - P)$	$-KP \ln P$	$3KP^{2/3}(1 - P^{1/3})$
Time interval to complete a specified amount of growth, P_0	$\dfrac{1}{K} \ln\left[\dfrac{P(1 - P_0)}{P_0(1 - P)}\right]$	$\dfrac{1}{K} \ln\left(\dfrac{\ln P_0}{\ln P}\right)$	$\dfrac{1}{K} \ln\left(\dfrac{1 - P_0^{1/3}}{1 - P^{1/3}}\right)$
Inflection point, P_i	$1/2$	$1/e$	$8/27$

a bottle, yeast cells growing in a fixed nutrient medium, and white rats in a finite space, namely,

$$P = \frac{P_0}{1 + A \exp{(-kt)}} \tag{1}$$

where P is the population or cell mass at time t, P_0 was the population at the beginning of the experiment, and A and k are parameters. This is identical to the logistic curve proposed for technological progress by de Solla Price (see Chap. 6).

Lenz also suggests an analogy with bisexual reproduction [9]. The various notions of conception (or germination), embryo gestation, birth, infancy, adolescence, maturity, lifetime, etc., are familiar ones with obvious parallels. Lenz goes further and identifies the male parent with existing art, the female parent with the inventor (!), the male population with "total inventions disclosed minus obsolete inventions," etc. This seems to be a farfetched and probably unnecessary attempt to derive a phenomenological model along the lines of the Pearl formula.

The real value of a biological metaphor is that it identifies natural stages of development, which succeed each other in a natural order. The analyst is reminded that from gestation to maturity are involved many intervening processes which cannot be skipped—nor can they be accomplished overnight—regardless of motivations or incentives.

Of course the similarities are far from exact. The biological analogy carries with it the erroneous implication that each technology has "a life of its own," and that its evolution is governed by some internal dynamical law, with stages of (relatively) fixed length or proportions. In fact the transition from one stage to another may in some cases be so rapid as to be almost unnoticeable, and in others it may be very protracted or never occur at all. It may happen that a technology approaches maturity for a specialized market, e.g., scientific laboratories, military or space applications, much faster than it does for more general purposes. Moreover, change need not always result in progress in the "forward" direction: instances of regression have occurred, where a technology reverts to an earlier stage as a result of developments in other fields. One might cite the cases of Stirling's hot-air engine, the early mechanical analog computers of Babbage, or Maelzel's remarkable mechanical robots, all initiated in the early nineteenth century and then allowed to lapse—only to be revived in the twentieth century.

However, despite ambiguities, imprecision, and some misleading implications, the life-cycle metaphor appears to have a considerable degree of utility as regards the ordering or structuring of events, and one need not quarrel with its use on a restrained basis. It seems worth illustrating the notion with a specific example.

Controlled Fusion Power

Conception/germination 1930s–1940s	Began with Eddington (1930), from study of astrophysics. Theory of exothermic nuclear fusion reactions was essentially fully developed and the conditions for a chain reaction were calculated.
Birth 1954	The first thermonuclear explosion in 1954.
Infancy 1955–1970 (?)	Began with the inception of "Project Sherwood" by the AEC in the early fifties, a program specifically aimed at producing fusion reactions on a sustainable and controllable basis. The major problem at this stage is learning to contain a dense hot plasma of ionized deuterium in a magnetic field and to control or eliminate the so-called plasma instabilities.
Childhood 1970–1980 (?)	Will be initiated, most likely, by the first self-sustaining production of neutrons of unmistakably thermonuclear origin. The problem will now be to scale up to larger sizes, to design equipment which will operate over a period of time without constant maintenance by highly skilled scientists and which will produce more power than it consumes. There will be a good deal of emphasis on vacuum systems, large-size superconducting magnets, etc.

Adolescence 1980–1995 (?)	Will be initiated probably by a pilot plant which produces 50 kw or so of power (however uneconomically). Larger and larger models will follow, operating more and more efficiently. The first practical plant may well be in some out of the way place like Greenland or Antarctica, at the bottom of the ocean, or on the moon, where the economics are abnormal. By the end of the period of adolescence, costs should have dropped to the point where they are competitive with the cheapest "conventional" source of power—probably nuclear fission! Also during this period there will probably be heavy developmental research on fusion-powered propulsion systems for interplanetary rockets.
Maturity 1995 (?)	Will begin when fusion power plants can be justified on the basis of purely economic calculations, and/or when thermonuclear rockets are in operation.

It is worth noting that the various stages can be marked, without seriously straining our credulity, by specific and expectable events or signals which can be identified in advance, although their dates cannot.

Another familiar metaphor for the evolution of scientific research and technology has been expounded by Gerald Holton [11]. It is the notion that scientific research is akin to exploring "an ocean of truth" or, perhaps, mining a "lode" of interesting ideas. As a zeroth approximation, Holton assumes the pool (or lode) corresponding to a specific area is finite. He then notes that as the first explorers open up the new lode, news of the discovery spreads and a "gold rush" phenomenon occurs: many new investigators defect from their old fields, in search of greener pastures. After a few years interest dies away as progress becomes harder and harder to achieve, and the field may become dormant or move ahead very slowly until (perhaps) something happens to rekindle interest.   Figure 7.3 is a schematic representation of these characteristic phenomena [12].

To explain why science as a whole does not also die out (at least in the same time scale), Holton suggests that the zeroth approximation must be modified.   In fact, there are very many—

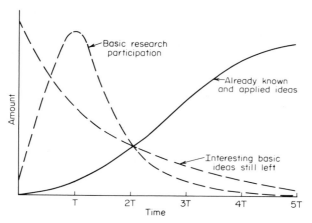

figure 7.3   *Scientific research and scholarship.*

perhaps infinitely many—lodes of interesting ideas.   New lodes of ideas are revealed when some fundamental discovery occurs which marks a branching point on the "tree of knowledge" (another metaphor illustrated by Fig. 5.7, which identifies some of the major new lodes which were sparked by nineteenth-century work on the propagation of sound waves in a compressible medium—air) [13].

Each new branch can, of course, branch again and again.   Thus the number of lodes being worked at any given moment tends to proliferate.   In fact, since the number of new branching points likely to be discovered in a given time is, at least roughly, proportional to the number of fields currently under active investigation, one is again indirectly back to an exponential law of growth.   However, progress in Holton's scheme does not occur uniformly or smoothly, but "stepwise," as the potential of each lode is exhausted, but other, newer lodes are opened up in succession. Holton has aptly termed this process *escalation.*[4]

Recently a program has been instituted by the American Academy of Arts and Sciences to explore historical analogies for great "social inventions." The particular example which has been chosen for study in depth is an analogy between the development of the railroads in the nineteenth century and the space program in the twentieth [14]. The editor, Bruce Mazlish, believes that the twentieth century is "the space age" by the same criteria that the nineteenth century was "the age of railroads." The fact that the analogy itself is a very questionable one[5] may help to explain why the study was relatively barren of insights regarding the likely social impact of a major invention. Mazlish ventures a number of broad generalities (although it is not clear how the study contributed to their formulation):

• All social inventions are complex rather than simple, and must be treated as such.
• No single social invention can be thought of as the unique economic "prime mover" of its epoch.
• All social inventions will have both positive and negative utilities.
• All social inventions develop in stages, and each stage has its own characteristic types of social impact.
• The effects of social inventions are indicated by "national style."

PHENOMENOLOGICAL MODELS

Despite the fact that several of the metaphors and analogies discussed in the previous section lead to explicit mathematical growth curves, the validity of these curves—particularly the Pearl formula—rests on thin and superficial resemblances between the fundamental processes involved. Increase of population and increase of knowledge are not the "same" in any deep or fundamental sense. The common use of a biological analogy in journalism

[4] A usage which has nothing in common with the modern terminology of diplomacy and warfare.
[5] The twentieth century can probably more convincingly be described as an age of electricity, an age of communication, an age of automation, or even an age of automobiles!

is evidence not of its underlying validity, but merely of its convenience and familiarity. The Pearl formula may be a phenomenological model of the growth of yeast cells in a bottle,[6] but it has only a poetic connection with the question of technological progress.

Ridenour (1951) was perhaps the first to look for mechanisms which would explain the behavior of technological trends [15]. He did not dig very deep: exponential increase is accepted without much analysis as a general empirical "law of social change," justified on the assumption that the rate of acceptance of a new product, service, or technique is proportional to the number of potential users (customers) who have been exposed to it, that is,

$$\frac{dN}{dt} = KN \qquad (2)$$

where K is the index of proportionality. However, Ridenour notes that the number of such potential users normally has a finite upper limit L. He suggests that K should not be a constant, but instead

$$K = k\left(1 - \frac{N}{L}\right) \qquad (3)$$

to approximate the likelihood that a person who is newly exposed to the new product, etc., happens to belong to the class of potential users. Evidently this probability declines as N approaches L.

When this expression for K is substituted into the first equation and integrated, the result turns out to be

$$N(t) = \frac{L}{1 + (L/N_0 - 1)\exp(-kt)} \qquad (4)$$

which is mathematically equivalent to the Pearl formula used earlier (although this is essentially accidental).

Recently there has been a tendency to view scientific progress in terms of information gain. For instance, Lawton Hartman proposes a basic relationship of the form [16]

$$\frac{dI(t)}{dt} = KNI \qquad (5)$$

In words, the rate of increase of information is proportional to the total amount of information which already exists, to a probabil-

---

[6] Although even in this case the model is far too simplistic.

ity K that a scientist encountering a "bit" (or unit) of information will "react" and create a new unit of information, and to the number of scientists N in the field.[7]   In a new field (e.g., computers, lasers) N may itself be exponential, for instance

$$N(t) = N_0 \exp k_1 t \tag{6}$$

whence

$$\frac{dI}{dt} = (KN_0 \exp k_1 t)I(t) \tag{7}$$

If K can be taken as a constant, this differential equation integrates to the double exponential form

$$I = I_0 \left[\exp \left(\frac{KN_0}{k_1} \exp k_1 t \right) - 1 \right] \tag{8}$$

However, if the total amount of pertinent information (like Holton's lode of possible ideas in a field) has an upper limit J—and for simplicity N is not changing—the probability of a reaction creating genuine new knowledge in the field might be assumed to decrease as the lode is exhausted.   In other words, K is given by the factor (seen earlier)

$$K = k \left(1 - \frac{I}{J} \right) \tag{9}$$

Assuming the above, but a constant unchanging N, the result of the integration is again the familiar logistic formula

$$I = \frac{J_0}{1 + (J/I_0 - 1) \exp (-kNt)} \tag{10}$$

The weaknesses and strengths of this model can be seen more clearly now that the contributory mechanisms are explicitly identified.   For instance, the model assumes that the probability of a "reaction" resulting in the creation of significant new information is proportional on the average to the amount of information encountered.   In other words, the better the technical library and

---

[7] Hartman used a number of concepts derived from the kinetic theory of gases (e.g., collision cross section, etc.), but the model is still a phenomenological treatment of information increase, subject to certain simplifying assumptions.

the more time the scientists spend in it, the more they should produce! This is misleading to say the least, since the researcher has only a limited amount of time and energy for all purposes; if he spends all of it reading the literature he will create nothing! Clearly a law of diminishing returns is at work which should be taken into account.

Moreover, the model presumes that no exogenous factors limit the addition of new knowledge to a field—i.e., that research is unrestricted and self-motivated. In actuality, there usually exists some kind of filter which screens the incipient ideas as they are originally produced by research workers, and discards the majority of them. The criterion of choice is rarely of scientific value: "Does it add new knowledge to the field?" The question usually is "Does it add to such and such a functional capability?" By following up only those ideas which contribute to a specific goal, of course, the rules of the game are distinctly modified. For one thing, significantly new knowledge cannot be thought of as something which is created instantaneously in a flash.[8] There is a gap between the time when the general nature of the future discovery is perceived by the researcher and the much later time when the result has actually been established and verified and published. In the interim, if the researcher is diverted to other tasks, the new information is wasted and cannot lead to further discoveries. Another hazard, of course, is that the information will be withheld deliberately for security reasons or to maintain a commercial advantage. Hence, as Jantsch has pointed out, the communication of information to the scientific community is quite imperfect due to high attrition between conception and publication. This is partly due to the strong mission orientation of many research organizations— which causes much potentially useful research to be discarded—and partly due to external pressures to limit free communication in certain fields.

Raymond Isenson suggests a slightly different basic model to describe the accumulation of knowledge, namely [17]

$$\frac{dI(t)}{dt} = K[N(t) + \lambda N^2(t)] \tag{11}$$

[8] Hartman's "collision" analogy.

where $I(t)$ is the stock of information at time t, and $N(t)$ is the number of scientists or research workers. Isenson's basic model differs from Hartman's in that the rate of increase is *not* proportional directly to the amount of information already in existence, although he assumes the same correction factor as Hartman as I approaches its upper limit J, namely

$$K = k \left(1 - \frac{I}{J} \right) \tag{12}$$

The quadratic term in $N(t)$ is an "interscientist communication factor," whose relative importance is measured by the size of the coefficient λ. A value of λ near $\frac{1}{2}$ would imply that the productivity of a group of scientists is more nearly related to the total number of possible communications links among them $[\frac{1}{2}N(N-1)]$ than their absolute number $[N]$. Of course, a small value of λ would imply that this effect is unimportant.

In the case where λ is equal to $\frac{1}{2}$ and $I \ll J$, the equation reduces to

$$\frac{dI(t)}{dt} \cong \frac{K}{2} N^2(t) \tag{13}$$

If $N(t)$ is given, this can be integrated. For instance, if

$$N = N_0 \exp (k_1 t) \tag{14}$$

then

$$I = \frac{K}{4K_1} N_0{}^2 [\exp (2k_1 t) - 1] \tag{15}$$

From the work of de Solla Price estimates for the factors K and K_1 in the Hartman and Isenson models can be made for the growth of scientific research as a whole [18]. If a "unit" of significant knowledge is approximately equivalent to a published paper, the productivity of the average scientist (K) turns out to be 0.1 units (papers) per year, but 0.8 units per year for the 90th percentile in productivity and 2.0 units per year for the 98th percentile. The constant K_1 can be deduced from the historical fact that the number of scientists doubles every 15 years, whence the annual rate of increase is

$$K_1 = \frac{\ln 2}{15} \cong 0.046 \qquad \text{(or 4.6\%)}$$

The major weaknesses of the Hartman model apply equally to

the Isenson model. When the number of scientists N is large, the quadratic term in Isenson's model has very little practical meaning, and it would probably be better to rely on the assumption that creativity is proportional to the total amount of knowledge in the field; where N is small, however, Isenson's model may be a better representation of the synergisms among an active "in group" [19].

A more sophisticated phenomenological model has recently been developed by Acey Floyd of Lockheed Aircraft Corporation [20]. It begins, in the spirit of morphological analysis, by supposing that, for a given value f of some figure of merit, there exist X "techniques" (including inventions) which would lead to an increase in f, as compared with a total of M possible techniques which could be considered. The a priori probability of "success" per attempt is

$$P(f,1) = \frac{X}{M} \tag{16}$$

If there are W research workers, each with a productivity N (attempts per unit time), the probability of success in time Δt is unity minus the cumulative probability of failure, or

$$P(f, \Delta t) = 1 - \left(1 - \frac{X}{M}\right)^{NW\Delta t} \tag{17}$$

To evaluate this expression it is convenient to introduce a relationship between the value of the figure of merit f and the number X of techniques which might improve it further. As f approaches its physical upper limit value F, X gradually decreases to zero as avenues of possible improvements are exhausted. Floyd chooses a simple form for this functional relationship, based on an analogy with absorption phenomena: the incremental decrease in X needed to achieve an incremental increase in f is assumed to be proportional to the number of techniques already "absorbed" in f, namely

$$\frac{\Delta X}{\Delta f} = -K(M - x) \tag{18}$$

This can be replaced by a differential and integrated

$$K \int_f^F df = K(F - f) = - \int_x^0 \frac{dX'}{M - X'}$$
$$= - \ln (1 - X/M) \tag{19}$$

whence

$$1 - \frac{X}{M} = \exp\left[-K(F - f)\right] \tag{20}$$

Substituting in the earlier expression for $P(f, \Delta t)$, we obtain

$$P(f, \Delta_i t) = 1 - \exp\left[-(F - f)K_i(t)N_i(t)W_i(t)\, \Delta_i t\right] \tag{21}$$

It can be shown that the cumulative probability of exceeding a value f is unity minus the cumulative probability of not exceeding f up to time t; the latter is the product of many similar terms for all increments of time $\Delta_i t$. This product is obtained by summing over the index i in the exponent, namely

$$\begin{aligned} P(f,t) &= 1 - \exp\left[-(F - f) \sum_i K_i N_i W_i\, \Delta_i t\right] \\ &= 1 - \exp\left[-(F - f) \int_{-\infty}^{t} KNW\, dt\right] \end{aligned} \tag{22}$$

where, again, the sum is replaced by an integral (although f is also a function of time, it is not included in the integration, because its value is considered *fixed* in the probability computation). It is easy to believe that K and N change relatively slowly and smoothly with respect to time; however W (the number of research workers in a field) may not only change rapidly, but very likely depends on the rate of progress in the field. Floyd suggests the following first order relationship to take this dependence into account explicitly:

$$W = W_0(t)(f - f_c)^p \tag{23}$$

where f_c is the figure of merit of some competitive technology. Thus the greater the difference between f and f_c the more workers tend to be attracted into the field ("bandwagon" effect). The final expression is

$$P(f,t) = 1 - \exp\left[-(F - f) \int_{-\infty}^{t} (f - f_c)^p T(t)\, dt\right] \tag{24}$$

where $T(t)$ is a composite, slowly varying function of time

$$T(t) = K(t)N(t)W_0(t) \tag{25}$$

Floyd has demonstrated that the time behavior of the figure of merit $F(t)$ can be determined in terms only of the function

$$\int_{-\infty}^{t} \frac{T(t') \, dt'}{\ln 2} = g(t) \tag{26}$$

The trick is to specify a particular probability of success—say 50 percent—and observe that for $P(f,t) = 0.5$ the bracketed exponents must have a constant and fixed value, namely $\ln \frac{1}{2}$ $(= -\ln 2)$, whence

$$\ln \frac{1}{2} = -(F - f) \int_{-\infty}^{t} (f - f_c)^p T(t') \, dt' \tag{27}$$

or

$$\frac{1}{F - f} = \frac{1}{\ln 2} \int_{-\infty}^{t} (f - f_c)^p T(t') \, dt' \tag{28}$$

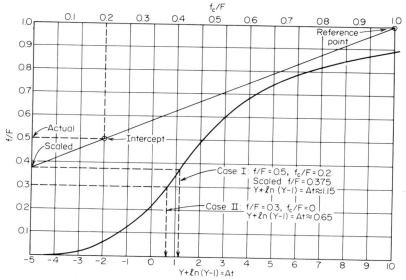

Rules: 1. Plot intercept f/F and f_c/F
2. Draw straight line from reference point (f_c/F = 1.0)
3. Determine scaled f/F
4. Read Y+ℓn(Y-1) on abscissa to scaled f/F

figure 7.4   *Nomogram for calculating $Y + \ln (Y - 1)$ when f/F and f_c/F are given* $\left(Y = \dfrac{1 - f_c/F}{1 - f/F} \right)$.

Differentiating both sides with respect to t one obtains

$$\frac{df/dt}{(F-f)^2} = (f - f_c)^p \frac{T(t)}{\ln 2} \qquad (29)$$

whence $\displaystyle\int^f \frac{df}{(F-f)^2(f-f_c)^p} = \int^t \frac{T(t)}{\ln 2} = g(t) \qquad (30)$

Making the substitution $Y = (F - f_c)/(F - f)$, the integral on the left has a simple closed form for the case $p = 1$, yielding

$$Y + \ln (Y - 1) + C = (F - f_c)^2 g(t) \qquad (31)$$

This expression can be inverted numerically to find Y (and therefore f) as a function of $g(t)$. Floyd made the approximation that $g(t)$ can be assumed to be linear, since $T(t)$ was "slowly varying." That is

$$(F - f_c)^2 g(t) = At + C \qquad (32)$$

Based on this assumption he has computed a nomogram from which f can be determined, as shown in Fig. 7.4. Any trend can be projected along the 50 percent probability trajectory once

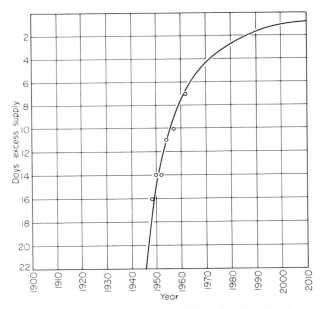

figure 7.5 *Days' supply of available stocks of crude oil in the United States.*

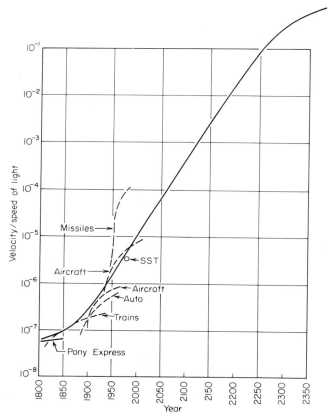

figure 7.6 *Speed trend (mechanical powered vehicles).*

f_c and F, plus two datum points to determine the two constants A and C, are known. Floyd has obtained impressively close fits for several sets of data, using his method. These are reproduced as Figs. 7.5 to 7.7.

One may be tempted to quarrel with Floyd's assumed form for $g(t)$, since the evidence suggests that $T(t)$—which encompasses changes in scientific productivity and in the numbered research workers—should probably be exponential. Thus $g(t)$ would also be exponential in form. However, if $T(t)$ is truly a "slowly varying" function as postulated, then $g(t)$ will change even more slowly, since it represents the integral over $T(t)$ starting from

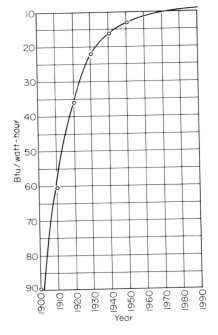

figure 7.7 *Efficiency of fuel-burning electric power plants.*

$t = -\infty$.   In this case, even though $g(t)$ is actually exponential, the linear approximation (32) should be satisfactory.

OPERATIONAL MODELS AND SIMULATION

The term "operational" has no epistemological significance: it merely connotes the use of operations-research methods, including linear programming, in large-scale, multivariate models designed to simulate the behavior of a very complex system or activity, such as a national or regional economy or the outcome of a nuclear war.   Operational models are conceptually, but not practically, distinguishable from the elaborate computer programs, languages, and data banks with which they are associated.   An example *par excellence* of this type of model is PARM,[9] a simulation system for providing guidance to the Director of the Office of Emergency Planning in the Executive Office of the President on resource allo-

---

[9] *Program Analysis for Resource Management.*

cations and the coordination of economic mobilization plans among the major agencies of the United States government. The model was designed by Marshall K. Wood and programmed by a group under his direction at the National Planning Association [21]. It involves keeping track of 1,007 distinct "activities," (or sectors) and approximately 94,000 "factor records"—or time-dependent coefficients—describing the relationships among the activities. PARM is essentially a very sophisticated input-output model of the Leontief type [22]. Of course, there are in existence many other examples of economic simulations, as well as war-gaming [23],[10] damage-assessment, consumer-behavior, voter-behavior, transportation, land-use, urban-planning, water-resources-planning, and other models.[11] A brief survey of some 37 computerized operations models is included among the studies prepared for the National Commission on Technology, Automation and Economic Progress in 1966 [24].

There have as yet been few attempts to simulate technological progress per se, and none, to the author's knowledge, have yet been implemented.[12] One early manual model described by Ralph Lenz [25] is based upon Jay Forrester's "industrial dynamics" approach, which is essentially a macroeconomic simulation of corporate behavior [26]. The variables involved run the gamut over sales, manpower (subdivided among categories, if necessary), cash flow, backlog, inventory, capital investment, etc. Lenz has focused on what he calls the "knowledge-progress system," relating the output of technology, measured in terms of productivity or systems performance, to various educational (knowledge) inputs. Some of the variables include

1. Total population eligible for training (ages eighteen to twenty-one)

2. Number of engineers and scientists in training

---

[10] In fact, there exists a Joint War Games Agency, JWGA, within the Pentagon, charged with the responsibility of developing and carrying on gaming and simulation activities for the Joint Chiefs of Staff.

[11] However structured-manual or computer-assisted "games"—e.g., to explore conflict situations—are not properly classed as models or simulations; rather, they are adjuncts to intuitive thinking (see Chap. 8).

[12] That is, programmed and run on a computer with "real" data.

3. Division of resulting manpower skills among research, teaching, and other technical occupations

4. Capacity of available research facilities to utilize manpower

5. Ultimate output of "knowledge and progress"

The "flow" from each stage to the next is governed by various feedback loops and exogenous conditions (such as research and development funds). About 19 empirical relationships and 37 proportionality factors were derived from an examination of past performance of the "system."[13]

Jantsch reports that Xerox Corporation has developed a complete business model involving 500 program variables as an adjunct to its corporate technological forecasting activity [27]. IBM Typewriter Div., Lockheed, Hercules Powder, ICI, and other corporations are also experimenting with the use of models of their own operations. Quantum Science Corporation is reportedly developing an elaborate input-output model of the United States electronics industry (MAPTEK) for use in forecasting the 5-year performance (i.e., sales, profitability, etc.) of this sector [28]. Two hundred equipment categories are to be included, with a further breakdown by circuit function and component (800 categories); fabricated materials and raw materials are to be added.

The attractive feature of the input-output approach is that irregularities in the trend curves show up explicitly, often in terms of readily identifiable causes such as "bottlenecks" in the number of teachers or of essential experimental facilities in a new field: thus superconductivity cannot be investigated on any reasonable scale without a helium liquefier for the laboratory. The unattractive feature is simply that in many cases the amount of quantitative apparatus required is too great in relation to the reliability of any conclusions one can reasonably expect to arrive at. The cumulative quantitative uncertainties due to combining and manipulating a large number of equations and proportionality constants, each one of which is likely to be individually somewhat imprecise, must be such as to make the net result often little better than a guess

---

[13] A typical example: the number of individuals desiring to enter technical training is roughly equal to an empirical constant times the eligible student population times the number of teachers.

(and worse than some guesses), except for short-term forecasts where the empirical relationships are close to the data points before "taking off" into new territory and the parameters can be forced to reproduce the historical time series. It is hard to take such a model seriously as a long-range forecasting tool at the present state of our knowledge, although it may be extremely useful as a teaching device for forecasters and decision makers, and as a contribution to the general methodological discussion.

REFERENCES

1. R. C. Lenz, Jr., "Technological Forecasting," ASD-TDR-62-414, Aeronautical Systems Division, Air Force Systems Command, June, 1962. (DDC Accession number AD 408,085.)
2. *Ibid.*
3. Henry Adams, *The Education of Henry Adams,* reprinted by Heritage Press, New York, 1958.
4. Erich Jantsch, *Technological Forecasting in Perspective,* OECD, Paris, 1967.
5. Lenz, *op. cit.*
6. Raymond Pearl and L. J. Reed, *Proceedings of the National Academy of Science,* vol. 6, p. 275, 1920; also, R. Pearl, *The Biology of Population Growth,* Alfred A. Knopf, Inc., New York, 1925.
7. A. J. Lotka, *Elements of Mathematical Biology,* Dover Publications, Inc., New York, 1956.
8. R. E. Ricklefs, "A Graphical Method of Fitting Equations to Growth Curves," *Ecology,* vol. 48, pp. 978–983, 1967.
9. Lenz. *op. cit.*
10. D. G. Samaras, "Nuclear Space Propulsion: A Historic Necessity," *Nuclear Energy,* London, September, 1962.
11. Gerald Holton, "Scientific Research and Scholarship: Notes toward the Design of Proper Scales," *Daedalus,* 1962.
12. *Ibid.*
13. *Ibid.*
14. Bruce Mazlish (ed.), *The Railroad and the Space Program: An Exploration in Historical Analogy,* Technology, Space and Society Series, American Academy of Arts and Sciences, M.I.T. Press, Cambridge, 1965.
15. Louis Ridenour, "Bibliography in an Age of Science," *2d Annual Windsor Lectures,* The University of Illinois Press, Urbana, 1951.
16. Lawton Hartman, "Technological Forecasting," in G. A. Steiner and Warren Cannon (eds.), *Multinational Corporate Planning,* Crowell-Collier Publishing Co., New York, 1966 (cited by Jantsch, *op. cit.*).
17. Raymond Isenson, "Technological Forecasting: A Planning Tool," in *Multinational Corporate Planning* (see ref. 16). (Cited by Jantsch, *op. cit.*) Also see *Symposium on Long Range Forecasting and Planning,* USAF Office of Scientific Research, AF Academy, August, 1966.

18. Derek de Solla Price, *Little Science, Big Science,* Columbia University Press, New York, 1966 (cited by Jantsch, *op. cit.*).

19. James D. Watson, *The Double Helix,* Atheneum Publishers, New York, 1968.

20. Acey Floyd, "Trend Forecasting: A Methodology for Figures of Merit," in J. Bright (ed.), *1st Annual Technology and Management Conference,* Prentice-Hall, Inc., Englewood Cliffs, N.J., 1968.

21. Marshall K. Wood, "PARM: An Economic Programming Model," *Management Science,* vol. 2, pp. 619–680, May, 1965.

22. Wassily Leontief, "The Structure of the U.S. Economy," *Scientific American,* April, 1965.

23. A number of such models are discussed in R. U. Ayres, "Models of the Post-attack Economy," HI-647-RR, and "On Damage Assessment Models," HI-693-RR, Hudson Institute, New York, 1966.

24. Abt Associates, Inc., "Survey of the State of the Art: Social, Political and Economic Models and Simulations," in Appendix vol. V of *Technology and the American Economy,* Report of the National Commission on Technology, Automation and Economic Progress, Washington, D.C., February, 1966.

25. R. C. Lenz, "Technological Forecasting," in *Long Range Forecasting and Planning,* Symposium at U.S. Air Force Academy, August, 1966.

26. Jay W. Forrester, *Industrial Dynamics,* The M.I.T. Press, Cambridge, Mass., 1961.

27. Jantsch, *op. cit.*

28. *Ibid.*

8 INTUITIVE METHODS OF FORECASTING

The title of this chapter is, at first glance, contradictory. Intuition is—according to the dictionary—"the acquisition of knowledge without resort to the powers of reason," which seems to deny the possibility of applying any method. In actuality, of course, those mental processes which are often termed intuition, "hunch," or "flash of genius" may in fact be descriptive of a sort of gestalt: a synoptic grasp of many interrelated aspects of a complex problem, forming an integrated whole which is much greater than the sum of its parts. In short, although intuition is an undisciplined and unreliable form of cognition, it is not—as some would have it—the antithesis of rational analysis, nor should we imagine that intuition cannot be improved by the application of system and order.

The simplest form of intuitive forecasting—the simplest of all forecasting methods—is the undiluted vision of a single prophet or seer. Its modern equivalent is what Ralph Lenz calls "genius forecasting" [1]. If a certified forecasting genius is available, of course, there is nothing more to be said. If, however, either the genius or the certification is in question—which is normal—the modern tendency is to rely instead on an opinion poll or a panel of experts. The use of either device will presumably wash out, by simple averaging, if not by more creative interactions among the panel members, the effects of individual bias, idiosyncratic hobbyhorses, and selective ignorance.[1] Ideally, members of a panel will utilize each other's special knowledge and achieve an interdisciplinary consensus. The consensus idea is not a foolish one, in principle, even though one remains uncomfortably aware of the fact that the future will be insensible to majority vote, no matter how distinguished and sophisticated the voters.

However, to the extent that the desired synergism does *not* occur—possibly because the experts are all busy in the world of affairs and have not the time, interest, or patience to stimulate, or be stimulated by, each other—the forecast is still essentially an opinion poll taken among a selected group. It is possibly relevant to note that the National Bureau of Economic Research has recently completed a 10-year study of economic forecasters, which unequivocally concludes that individuals or small working groups do consistently *better* than consensus polls, at least on economic issues [2].

A possible reason for this is that the poll is a conservative, informal average, weighted a little bit by intensity of conviction (if the experts talk to each other on a panel and modify their forecasts accordingly). However, the most nearly correct forecast is very often the most radical one, as Clarke and others have frequently pointed out [3]. Thus the first major forecasting effort by a panel, the 1937 study by the Natural Resources Committee of the National Research Council [4], was a very sober and responsible document which missed virtually all of the major developments of its decade, including antibiotics, radar, jet engines, and atomic energy (see Chap. 1). The ludicrous failure of a blue-ribbon committee

---

[1] For instance, F. W. Lindemann (Lord Cherwell), Churchill's powerful and stubborn scientific advisor during World War II, advocated solid-fuel rocket propellants so single-mindedly that he refused to believe that the German V-2 rocket could fly, despite photographic evidence to the contrary. He later told the House of Lords that the ICBM was not feasible (see Chap. 1).

set up to evaluate the prospects for jet engines in 1940 was also cited in Chap. 2.   Evidently overconservatism is a natural characteristic of committees.[2]

Another problem with the committee-of-experts approach is that the "consumer" of a committee report often has difficulty knowing whether the group interacted in practice as it ought to do in theory; a clear advantage of a more explicit methodology might be that the underlying reasons for many conclusions could be brought out into the open and critiqued more adequately than is frequently the case—especially where the members of the committee are chosen for their standing.

STRUCTURED MAN-MAN INTERACTIONS

Since committees of the conventional type are notoriously conservative, one of the obvious approaches is to deliberately stimulate radical, "way out" thinking.   The best-known procedure is *brainstorming,* which consists of group meetings conducted under a set of simple rules designed to create an environment conducive to free-wheeling speculation, such as the following:

1. Focus on a single well-defined problem—but
2. Consider any idea, regardless of apparent relevance or feasibility
3. Do not criticize any idea
4. Do not explore the implications of any idea

A trained group leader is usually required to make a brainstorming session work.   His job is simply to referee, remind the participants of the rules, and otherwise encourage uninhibited discussion. There are a number of variants, depending on the purpose of the activity [5].

An increasingly important technique is *gaming,* in which each participant is asked to play (i.e., simulate) a specific role in a

---

[2] All revolutionaries are aware of the Leninist warning that the most dangerous threat to its leadership always comes from the radical left.   On technical committees that unspoken rule is generally to avoid being outflanked on the right (i.e., by the more conservative elements)!

scenario. The initial conditions only may be specified in advance, or an outline of the further action may be provided. The participants may be allowed to play their parts quite freely, or under severe constraints (the rules of the game), depending on the purpose of the exercise. This approach is already quite highly developed in the area of high military-political strategy,[3] where it is used often for training purposes and sometimes as a source of strategic or tactical insights regarding future environments and requirements. This application will be discussed in the next chapter, on planning. Role-playing "games" are also being introduced in a variety of other contexts, including industrial management [6], career and educational counseling, and even race relations [7].

A scenario is a logical and plausible (but not necessarily probable) set of events, both serial and simultaneous, with careful attention to timing and correlations wherever the latter are salient. In scenario writing one major emphasis is usually on the critical branch points (or saddle points), where small influences may have great effects on the outcome (e.g., of the game). It is sometimes illuminating to show explicitly how a single archetype scenario can generate families of variants as elements are changed. Perhaps the most authoritative discussion of the purposes and values of scenario writing may be found in the writings of Herman Kahn [8–11]. Kahn cites two advantages worth noting:

1. Scenarios are an effective tool to counteract "carry-over" thinking, and to force the analyst (or policy maker) to look at cases other than the straightforward "surprise-free projections."[4]

2. Scenarios are an antidote for concentrating exclusively on the forest and ignoring the trees: analysts who limit themselves to abstract generalizations may easily overlook crucial details and dynamics (because no single set seems especially worthy of attention), even though looking at some random specific cases can be quite helpful.

The scenarios which result from a serious gaming session (like

[3] Notably by the Joint War Games Agency in the Pentagon and by certain "think tanks" such as RAND Corp., Stanford Research Institute, and Hudson Institute.
[4] Also a Kahn term [10].

scenarios created "out of whole cloth" by analysts) may have value in studying *future environments* which, in turn, ultimately help determine technological requirements.   Thus, in the Department of Defense the study of future potential threats (conflict environments) is a major activity involving annual expenditures in the millions of dollars.   Similar scenario studies are carried out from time to time by NATO, the Arms Control and Disarmament Agency, and other agencies.   Although consideration of the potential impact of technological breakthroughs on the future military-political-economic-social environment—via gaming or scenario writing—is not currently a major component of the above-mentioned activities, it has been explicitly incorporated in certain special-purpose exercises.

The Boeing Company has developed a gaming procedure called "dynamic contextual analysis" to explore future military/space requirements and opportunities.   This procedure is fairly typical of other structural interactions involving role playing.   In this case a study group is divided into "red" and "blue" teams representing adversaries in a "cold war" situation.   Each team prepares (or is given) a political-economic philosophy and a set of basic national goals consistent with it.   These are translated into a set of political, economic, and military policies and strategies.   A referee or "control team" provides a particular scenario or "world future" to be explored, usually in the form of a brief summary of the events leading up to and the general situation prevailing at the starting point of the game.   Each team then chooses a series of actions in support of its stated goals and objectives.   These are monitored by the "control" and, if approved, transmitted to the opposing team.   Teams move in turns until their objectives are achieved or "general war" breaks out, or until the game is halted by the referee for some reason.   Proposed actions by each team are judged by the control team on the basis of relevance, credibility, and feasibility with respect to (1) the team's objectives and policies, (2) the international political situation, (3) the national economy, (4) national security (military posture), and (5) technological feasibility as a function of time.

National policies in the game may range from peaceful coexistence to world domination.   Strategies used to implement these policies can include political subversion of uncommitted nations, economic aggression, sponsorship of "wars of national liberation," direct military attack, disarmament and arms-control agreements, cooperative space exploration, military exploitation of cislunar space and/or the moon, etc. Many forms of tactics in support of the above strategies are permitted in the game, including subsidies to revolutionary or status quo groups, diversions of resources, military buildups and implied threats in strategic, "conventional," or new weapons such as biological or chemical warfare (BW or CW), orbiting bomb, extended antisubmarine (ASW) capability, etc.

In a series of nine actual simulations carried out by representatives of Boeing, Bendix Corp., Minneapolis-Honeywell, and Thiokol Chemical Corp. many actions were undertaken by red and blue—either provocative or in response to provocation—with technological implications, but only available technology and fairly straightforward extrapolations thereof were allowed. However, possible functional breakthroughs were systematically examined in a special round table session, not for feasibility but for their military and economic implications. The results of the latter exercise were invariably interesting and—in some cases—contrary to popular notions. For instance controlled fusion power was found to be an incremental improvement over existing capabilities, whereas the ability to broadcast power over great distance would offer several order-of-magnitude increases in capabilities for utilizing space, transportation, and weapons (e.g., laser ABM).

An interesting and compact "Future Game" stressing new technological capabilities which can be played by laymen has been developed by Theodore J. Gordon, Olaf Helmer, and Hans Goldschmidt and packaged and distributed (for public relations purposes) by the Kaiser Aluminum and Chemical Co.

Another form of structured interaction among a group has been developed by Olaf Helmer et al. [13,14]. As applied to the problem of forecasting it has become known as the *Delphi* method. The face-to-face interaction among members of a panel is eliminated. Instead of group discussion, or serial presentation of "position papers" to the group, the experts are interrogated individually (usually by questionnaire) with regard to their expectations for a series of hypothetical future events. A typical question in a Delphi study might be: in what year (if ever) do you expect electric cars to capture 10 percent of the automobile markets?

After the first set of questionnaires is obtained, the numerical answers are assembled as distributions, stated in terms of means and quartiles, plus any pertinent comments by the experts. Thus, at the beginning of the second round the participants might be told that the mean fell on 1983, with 25 percent of the estimates prior to 1978 and 25 percent after the year 1995, and would also be presented a list of anonymous comments and arguments for various positions, including some additional questions by the interrogators. Respondents are then asked to submit revised estimates together with reasons for agreeing or disagreeing with the initial consensus. In the third (and later rounds) this procedure is repeated, with additional commentary and (impersonal) debate. Eventually, if the method works, the respondents will articulate and clarify their thinking and the most convincing arguments will "win," with a resulting convergence and narrowing of the range of estimates. Such a convergence did, in fact, occur in an ambitious 1964 study reported by T. J. Gordon and Olaf Helmer [15]. The pattern has been repeated in a number of other Delphi studies carried out since then.

There is no doubt that the Delphi method eliminates some of the major objections to the use of committees, which arise largely from psychological factors such as unwillingness to back down from the publicly announced positions, personal antipathy to or excessive respect for the opinions of a particular individual, skill in verbal debate, bandwagon effects, etc. Since the method requires public opinion sampling techniques, it may, of course, introduce other kinds of bias. (What is the effect of the self-selection principle among respondents? It would be interesting to see how the nonrespondents *would* have responded if they had been persuaded to do so.) It is also difficult to allow for the bias of the pollster: it is a safe guess that the framers of the questions can to some degree guide the trend of the answers. It is a safe guess, also, that on being told he differs from the consensus by a large factor, and asked to reconsider, a respondent will either shift toward the consensus or stand firm; he is most unlikely to shift in the other direction. Moreover, in eliminating the group interactions, the implicit weight factors which committee members soon attach to each other's opinions—giving least weight (on a given issue) to the opinion of the least informed—is replaced by an assumption of self-selection, with an a priori equal weight given to all respondents on all issues.

As noted above, the distribution of individual forecasts tends to become progressively narrower and more sharply peaked as the successive rounds of interrogation and commentary occur. Thus it is clear that some group interaction is, in fact, taking place. What is not yet clear is how much of the convergence achieved in a typical group is the result of thoughtful consensus and how much is merely a disguised tendency of the less-convinced to adjust differences with the more-convinced. Specious persuasion is not necessarily eliminated by impersonalizing the interaction; indeed, the contrary may be true, since many common fallacies are superficially plausible and require deeper analysis to penetrate than the average busy expert may be willing to expend on answering a questionnaire.

Some of the strengths and weaknesses of the Delphi method can be illustrated with the help of an example from the study cited above. Figure 8.1 shows a world population projection based on "naive" trend extrapolation and, beneath it, an envelope of projections obtained from a panel of experts interrogated separately by Gordon and Helmer. The experts, in this case, clearly took account of the existing trend, but almost unanimously agreed that the future rate of population growth would slow down for various reasons, notably the continuing invention and wider acceptance of birth control measures. The panel was also cognizant of contrary trends, such as a continued decline in the death rate due to medical progress, and the possibility (but not probability) of greater improvement in the technology of food production and distribution. The result is a balanced forecast in which the best information available has been utilized in a way that no simple model or statistical extrapolation could hope to duplicate.

On the question of food, however, one finds that the panel concerned with "science" concurred in the prediction that large-scale "ocean farming" will be practiced in the early part of the twenty-

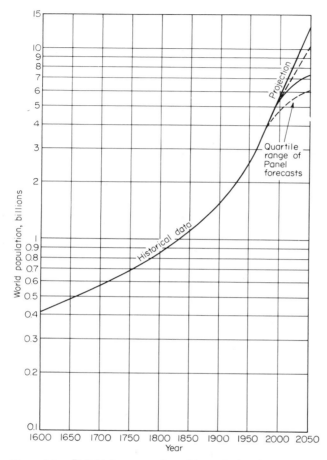

figure 8.1 *Delphi forecasts of world population (1960–2000)*.

first century. A systematic look at the planetary food supply suggests, however, that there are very good reasons for skepticism [16].

1. The maximum potential output of the ocean, even if farmed efficiently, is less than that of the land, but the difficulties of "seeding," "weeding," and "breeding" in the ocean are immensely greater.

2. More promising alternative sources exist, notably via microbial "digestion" and transformation of cellulose into protein.

Cellulose and lignins are currently being wasted in immense quantities as lumbering and paper-making residues, bagasse, cornstalks, wastepaper, trash and wood scrap, garbage, and sewage. Hydrocarbons may also be utilized in this fashion.

3. In any case, ocean farming requires either massive international cooperation or allocation of territorial rights in the ocean to particular nations, organizations, or individuals. This would imply a drastic change in the historical tradition of "freedom of the sea" and the law of capture which currently applies to fish and other marine life. Such a change would be nearly impossible to achieve international agreement on in the next 50 years.

If these arguments are correct, then the panel must have been wrong, and it may be useful to list several possible contributory factors:

1. At the time of the study oceanography was in vogue, and members of the panel may have been influenced by some of the widespread speculation.

2. It is widely, but probably falsely, assumed that the ocean is potentially much more productive than the land [17]. Panel members—who covered all the sciences, but were concentrated in the physical sciences and engineering specialties—may have been largely ignorant of this. Many nonbiologists also probably underestimate the ecological complexity of the ocean and the technological difficulties of "farming" in a three-dimensional medium where the photosynthesis is being carried on by microscopic organisms and many of the "grazing animals" are also microscopic.

3. The panel evidently assumed the problem is basically technological and that, once other food sources were exhausted, the necessary scientific resources would be mobilized to "solve" it. Many people can be forgiven for assuming that almost any scientific-technological problem can be solved within 50 years. However, the constraints are not all physical: the critical ones are probably political and legal. Since a necessary (but not sufficient) condition for ocean farming is a radical international reorientation in these spheres, it is obviously essential to consult experts in the relevant fields. It is a reasonable guess that nobody on the "science" panel was even slightly qualified in this respect. Obviously the panel might have considered the points made above and still

come to the same conclusion, for reasons of its own. Or it may have been guilty of overoptimism and superficiality, as charged. In either event, this is not an indictment of the Delphi method per se, but merely a scenario illustrating how and why it might go astray despite its built-in error-detecting and correcting features.

A modified version of the Delphi technique has been developed by Harper Q. North and Donald L. Pyke for internal use by the Thompson-Ramo-Wooldridge Corp. (TRW) [18]. In the first version, called Probe I, 27 senior technical experts from different divisions and departments were polled anonymously and asked to anticipate major technical "events" having a major potential impact on TRW and which they expected to occur during the period 1966 to 1985. A list of 401 events were compiled and examined. One in particular, the electric automobile, seemed to offer both significant threats to one part of TRW's business (the Automotive Group) as well as major opportunities to the other three Groups (Systems, Equipment, Electronics). It was decided, therefore, to undertake a major in-house study of this area involving, among other things, a rather elaborate rerun of the original survey. To participate in Probe II, 140 respondents were chosen in such a way that each operating division of the company was represented if possible by an expert in each of 15 selected categories of topics in which the division has an interest [19]. Panel-members of Probe II were again asked to list important technological events, by category, together with three separable evaluations as follows:

1. Desirability (badly needed; desirable; undesirable but possible)

2. Feasibility (highly feasible; likely; unlikely but possible)

3. Timing (year in which probability of occurrence exceeds 20 percent; 50 percent; 90 percent)

In the second round each panelist is provided with a composite list representing all predicted events in his own category plus those of related categories, but not including the desirability/feasibility/timing evaluations supplied by the originator. Each panelist now evaluates all of the events with respect to the three factors above. Panel members are asked to pay particular attention to making their forecasts self-consistent with regard to the timing of

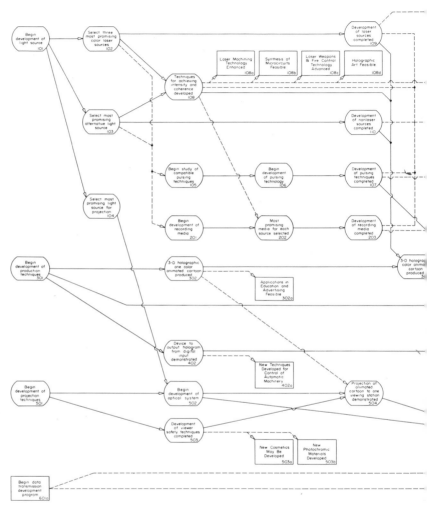

figure 8.2 *A program for the development of 3-D color holographic movies*
—SOON chart.

related events. In the third round, discrepancies and inconsisten-
cies are to be resolved on an individual basis. At the time of
writing only the first round is complete, with a list of about 1,750
events having been produced (to be somewhat reduced by edit-
ing). Final results of Probe II are scheduled to consist of 15

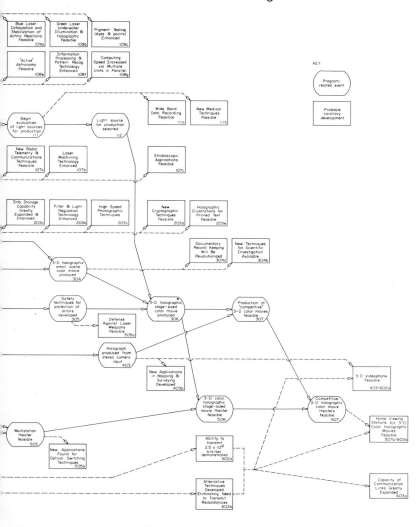

monographs summarizing the results of each category, plus an over-all document.

Probe II is designed to combine the Delphi methodology with analytical methods. Whenever future technological events important to TRW are identified by the panel, the intervening stages—

prerequisites and alternatives (if any)—are exhibited graphically in flow diagrams called SOON[5] charts.   Each intermediate technological development leading up to the main event, of course, has various corollary (by-product) implications which may offer substantial additional inducements to enter the field if they complement the company's existing activities.   The development and use of SOON charts is quite distinct from the use of Delphi techniques but, because of their close relationship in Probe, one chart depicting the development stages leading to holographic color movies from the North-Pyke study is shown as Fig. 8.2 [20].   As indicated on the chart the successful development of holographic color movies would also lead to many other useful techniques and devices.

**STRUCTURED MAN-MACHINE
INTERACTIONS**

An approach toward obtaining a meaningful consensus which achieves a greater degree of involvement at the cost of sacrificing some (but not all) of the benefits of anonymity is the structured man-machine "game."[6]   A simple version is essentially an automatic electronic voting machine which can be used by a group of M people seated simultaneously in an auditorium or around a table.   It consists of a set of M individual consoles (e.g., N-position rotary switches), one at each seat, which can be operated easily and inconspicuously.   At the front of the room, visible to all, is a display panel consisting of a matrix of M columns and N rows, one *column* for each seat having a console, and one *row* for each position of the rotary switch [21].   Thus each person has control of one column of the matrix and by turning his switch he can cause one of the N lights in that column to light up.   Each participant will quickly discover which column is his, but he has

---

[5] *Sequences of Opportunities, and Negatives.*

[6] Also sometimes (but inaccurately) called "models." Structured man-machine or man-machine interactions can also be used to forge a useful consensus for purposes other than direct forecasting, notably to determine criteria for deciding the (quantitative) relative importance of parallel goals or objectives of a program or policy. Such choices are an essential prerequisite of a rational resource-allocation or research-evaluation scheme (see Chap. 9).

no means of identifying the others, which will not be in any particular order *vis-à-vis* the seats.

The electronic voting machine can be used to carry out a telescoped Delphi study in a single sitting. Questions are asked by a leader, and votes are taken anonymously. Discussion may be verbal and direct, or it may be "filtered" in various ways to maintain the anonymity of the respondents.[7] With a little extra wiring the voting machine can also tabulate the results, and compute averages, means, quartiles, and even standard deviations.

From this rather simpleminded beginning, of course, the role of the voting-machine-*cum*-computer can be elaborated indefinitely. Many of the activities called for by the structuring of the games, for instance, can be performed rapidly and reliably by computers. Scenarios can be analyzed morphologically (see Chap. 5) and identified in shorthand fashion as combinations and permutations of their elements, under the allowed rules. Once the rules, the initial conditions, the actors and the actions are specified (from appropriate sets of allowed alternatives) the computers can run through the moves of the game internally[8] and "write" the resulting scenario. Since the number of possible game histories (scenarios) which can be created out of realistic numbers of players and components quickly becomes astronomical, the computer's assistance is invaluable. With the help of electronic data processing it becomes possible to "look" at hundreds or thousands of cases, whereas plodding human players might take several days to go through a single one.

[7] For instance, each person might speak into a throat mike and hear only through earphones. Voice characteristics can be altered and disguised electronically. Alternatively comments and questions can be passed directly to the leader in writing, and relayed by him.

[8] This is a purely mechanical "bookkeeping" procedure, as distinguished from a computer *simulation* (i.e., *model*) of a competitive game like chess. In the second case the moves are not picked at random out of an allowed set, but are determined at each stage by means of appropriate algorithms derived from some strategic theory (which is, of course, built in to the program by its designer). Incidentally, it is well within the bounds of possibility that a computer could learn to "play"—in the latter sense— a complex but finite, well-defined, *N*-person war game with explicit rules such as RAND's SAFE (*S*trategy *a*nd *F*orce *E*valuation) game thus carrying out a simulation of a simulation.

The so-called TEMPER[9] model, developed by Clark Abt Associates for the Joint War Games Agency in the Pentagon, is actually a rather sophisticated man-machine system designed to relieve the participating human "players" of a great deal of tedious computation and data manipulation and thereby increase their ability to zero in on the real issues being explored [23,24].

C. West Churchman, at the University of California, is apparently the first to develop a computer game related specifically to technological change. It is designed to simulate, with the assistance of role playing, the process of acceptance of and adaptation to new technology within a firm [25].

Man-machine systems such as the latter two are often classed as *operational models* although they are not used as such [26]. Rather, they are adjuncts to intuitive thought. However, regardless of what they are called, they are likely to play an increasingly important role in the future.

REFERENCES

1. Ralph C. Lenz, Jr., "Technological Forecasting," 2d ed., Report ASD-TDR-62-414, Aeronautical Systems Division, AF Systems Command, June, 1962. (DDC accession number AD-408-085.)
2. Victor Zarnowitz, "An Appraisal of Some Aggregative Short Term Forecasts" (Preliminary), National Bureau of Economic Research, December, 1964. Briefly summarized by *Business Week*, Jan. 9, 1965, pp. 39–40.
3. Arthur C. Clarke, *Profiles of the Future,* Harper & Row, Publishers, Incorporated, New York, 1963.
4. W. C. Ogburn et al., *Technological Trends and National Policy,* U.S. National Research Council, Natural Resources Committee, 1937.
5. For further discussion and references see Erich Jantsch, *Technological Forecasting in Perspective,* OECD, Paris, 1967.
6. *Ibid.*
7. See for instance, James S. Coleman, *Models of Change and Response to Uncertainties,* Prentice-Hall, Inc., Englewood Cliffs, N.J., 1964. Also a series of papers by Ithiel de Sola Pool and others associated with Simulmatics Corp., "Social Research with the Computer," special issue of *The American Behavioral Scientist,* vol. 8, no. 9, May, 1965.
8. Herman Kahn et al., "On Alternative World Futures: Issues and Themes," Report HI-525-D, Hudson Institute, New York, May, 1965.
9. Herman Kahn et al., *On Escalation: Metaphors and Scenarios,* Frederick A. Praeger, New York, 1965.

[9] *Technical Economic Military Political Evaluation Routine.*

10. Herman Kahn and Anthony J. Wiener, *The Year 2000: A Framework For Speculation,* The Macmillan Company, New York, 1967.
11. See also R. U. Ayres, "Methodology for Post-attack Research," HI-647-RR, Hudson Institute, New York, August, 1966, and an earlier study "Special Aspects of Environment Resulting from Various Kinds of Nuclear Wars, Part II: The Use of Scenarios for Evaluating Post-attack Disutilities," HI-303-RR, Hudson Institute, New York, Jan. 8, 1964.
12. "Military Space Requirements 1963–1973," Final Draft, the Advanced Military Space Requirements Council (representing the Boeing Co., Bendix Corp., Minneapolis-Honeywell Regulator Co., and Thiokol Chemical Corp.), December, 1963.
13. Olaf Helmer and N. Rescher, "Epistemology of the Inexact Sciences," *Management Science,* vol. 6, 1959.
14. Olaf Helmer, *Social Technology,* Basic Books, New York & London, 1966.
15. Theodore J. Gordon and Olaf Helmer, "Report on a Long-range Forecasting Study," RAND Corp., p. 2982, September, 1964.
16. R. U. Ayres, "Food," *Science Journal* (London), Special Issue on Forecasting and Future, October, 1967, p. 103. Also R. U. Ayres, "Technology and the Prospects for World Food Production," in Working Papers of the Commission on the Year 2000, vol. II, Appendix A, Hudson Institute, New York, 1967.
17. Schmidt, "Planetary Food Potential," *Proceedings of the New York Academy of Science.*
18. Harper Q. North and Donald L. Pyke, "Technology, the Chicken—Corporate Goals, the Egg," in J. Bright (ed.), *1st Annual Technology and Management Conference,* Prentice-Hall, Inc., Englewood Cliffs, N.J., 1968.
19. Harper Q. North and Donald L. Pyke, "Corporate Experience with Delphi," presented at *2d Annual Technology and Management Conference,* Washington, D.C., March 1968.
20. *Ibid.*
21. Such a machine has been built by Howard Wells of Bell Aerosystems, Inc.
22. Olaf Helmer, "How to Play SAFE: A Book of Rules for the Strategy and Force Evaluation Game," RM-2865-PR RAND Corp., November, 1961.
23. Abt Associates, "Survey of the State of the Art: Social, Political and Economic Models and Simulations," in *Technology and the American Economy,* Appendix vol. V, Report of the National Commission on Technology, Automation and Economic Progress, Washington, D.C., February, 1966.
24. Clark Abt, "War Gaming," *International Science and Technology,* August, 1964.
25. Abt Associates, "Survey," *op. cit.*
26. e.g., by Jantsch, *op. cit.*

9 POLICY AND STRATEGIC PLANNING

In Chaps. 5 through 8 the focus of our discussion has been explicitly on methods of forecasting technological change per se. Within this framework it is perhaps natural to ignore the dynamics of change and, for convenience, to treat the problem in an *ontological* context (recall Chap. 3), even though goals and needs actually play a dominant role—as has already been remarked. It is, therefore, appropriate and necessary now to consider the *teleological* (or *normative*) aspects of technological change: "inventing the future" in Dennis Gabor's neat phrase [1]. This will be the primary task of this and the following chapter.

In real life inventing the future is the province of planning and programming, although a vital prior role may be granted to the processes of recognizing or formulating long-range needs or goals at various levels of social, political, or economic organization. It is sometimes useful to visualize a hierarchy such as that in Table 9.1.

TABLE 9.1 A Hierarchy of Imperatives

Level	Essential purpose
Biological (needs)	To survive and reproduce the species
Individual (needs, wants)	To be happy, to be good (or avoid sin), to achieve personal status or power, etc.
National (United States) (historical purposes)	To perfect democracy and "freedom," to promote social justice and equal opportunity, to protect citizens, to defend "national interests," to provide the means for economic growth, etc.
Defense Department (goals/postures)	To deter attack and to fight wars
Navy (objectives, missions/strategies)	To protect sea lanes (transoceanic shipping routes) and control the oceans; to fight hostile naval forces and support United States land and air forces
Carrier Task Group (missions, tasks/tactics)	To carry out operations in support of the above.

Analogous distinctions between levels of analysis can be made for any type of organization or structured collection of functions and activities including a city, an industrial corporation, an industry, a labor union, the whole educational system, the monetary system, the Roman Catholic Church, the Salvation Army, the Republican party, or the American Medical Association. In every case needs, purposes, aims, goals, objectives, missions, and tasks[1] are listed in decreasing order of generality or universality, and increasing order of specificity. There is a parallel hierarchy of levels of means ranging through policies, postures, strategies, and tactics. These words are again not precise but they convey a sense of the same transition from the general to the specific.

This hierarchical representation will reappear several times in various guises later on; at this point it provides a natural framework for distinguishing several levels of planning in government or industry (or, indeed, any other context). The following definitions are adapted from Jantsch [2]:

[1] The logical order here is not supported by dictionary definitions of the words, which have overlapping meanings in any case. The usage is fairly common, however.

Policy planning: formulation of alternative goal patterns or functional objectives for the future—based on alternative future environments or scenarios—in a (continuous) comparison, selection, and feedback process. Policy planning—being concerned with goals—seldom involves technological considerations in any central way.

Strategic (or entrepreneurial) planning: formulation of a set of alternative routes or options for achieving the chosen set of goals, together with a procedure for systematic comparison and assessment. The result is what Jantsch calls a *decision agenda.* Frequently some—and sometimes all—of the strategic options involve significant technological developments, especially when the

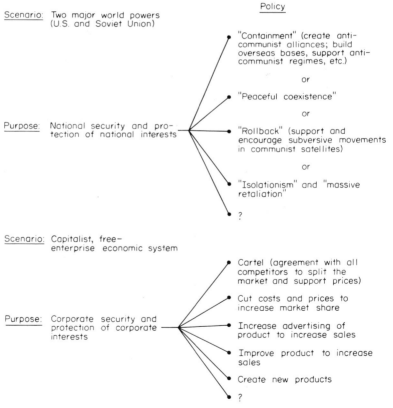

figure 9.1   *A menu of possible high-level strategies (policies).*

goals are related to physical (or biological) problems such as putting a man on the moon, feeding the world's population, cleaning up polluted rivers, or developing a defense against ballistic missiles. Nevertheless there may be a large number of ways to achieve any of these functional objectives. Figure 9.1 illustrates in a rather simplistic and schematic fashion the interrelationships between scenarios, purposes, and policies (or high-level strategies); Fig. 9.2 carries the same general theme down to the next lower level.

Tactical (or operational) planning: delineating the sequence of actions necessary to implement a particular strategy. The technological aspects of tactical planning would be concerned with reaching well-defined technological (as opposed to functional) objectives generally in terms of specified systems or subsystems.

Planning at the tactical level—what most people think of as "planning"—is discussed in the next chapter.

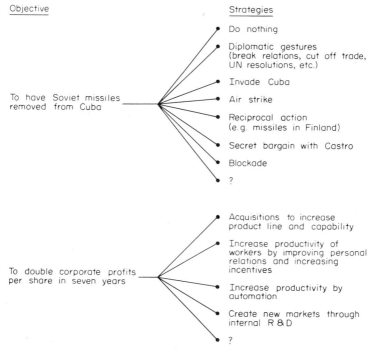

figure 9.2 *Strategies to achieve specific objectives.*

POLICY PLANNING METHODS

As stated earlier, policy planning consists of formulating and deciding among alternative sets of goals. Prior to formulation of goals, of course, it is necessary to reorganize problems. Many problems at the national or corporate level have only a peripheral connection—if any—to technology. However, when facing up to the implications of the population explosion, or the basic problem of defending the nation against nuclear attack, it is impossible to avoid certain broad technological questions such as "Are conventional resources inherently capable of supplying enough food to feed 6 billion people by the year 2000?" If the answer is yes, then one set of policies follows; if not, another set must be considered. Similarly, in a defense context it was at one point crucial to know whether it is technically feasible to explode nuclear weapons underground without the detonation being detectable at a distance by means of seismic instruments.[2]

As a rule, technological considerations at the policy level can be handled in a rather simplistic fashion (although there are obviously exceptions such as the foregoing example). Not infrequently the problems themselves are partly or even largely created by technology. Examples include the widening economic gap between rich and poor nations, the "brain drain," the domestic threat of unemployment arising from automation, the mounting problems of pollution, waste disposal, and despoliation of our physical and biological environment, substitution of synthetic organic products for natural raw materials produced in many cases by undeveloped monoagricultural nations,[3] and the "revolution of rising expecta-

---

[2] This question arose at the time of the controversy over the nuclear test ban treaty. Edward Teller, who opposed the treaty, propounded the so-called "de-coupling theory," which suggested that seismic disturbances from an underground nuclear explosion could be muffled if the weapon were exploded inside a large cavity. It was subsequently determined that the size of the cavity necessary to achieve this result would be so large as to be essentially infeasible in practice, but one result of the furor was that underground nuclear tests were removed from the scope of the treaty.

[3] A partial list of products subject to substitution or competition would include rubber, quinine, cotton, silk, wool, hemp, jute, copra, and (increasingly) sugar.

tions," which seems to be a by-product of the "information explosion" and modern innovations in rapid communications (recall the discussion in Chap. 4).

All of the above, and many other factors—including the very rapid change in weapons technology—contribute to a picture of the "future environment" within which functional objectives must be chosen by the decision makers. The environmental analysis phase of the process of policy choice (like other phases) is gradually becoming more explicit and more systematic than formerly. The primary decision method is still intuitive judgment based on experience, but refinements such as scenario writing, gaming, and Delphi (see Chap. 8) are coming into greater use, especially by "scientific" managers.

In the military establishment, environmental analysis is commonly known as *threat analysis*. It systematically considers a variety of possible political developments and other circumstances—based on intelligence inputs—which might lead to conflicts at various levels of violence, in various parts of the world. Possible United States responses to hypothetical outbreaks consistent with various basic national security policies (BNSP's) can be simulated by sophisticated gaming techniques (as described in Chap. 8), leading to the identification of potential "hot spots," vulnerabilities and weaknesses in existing contingency plans, and new or revised strategic concepts. The application of morphological analysis to generate scenarios was discussed in Chap. 5. At a subsequent point in the military planning cycle these overall strategic concepts are translated into military-capability requirements, force-development objectives, strategic-operations plans, and research and development objectives. Technological forecasting is deeply involved, but mostly at the strategic or tactical level of planning.

There are a large number of methodologies and procedures which can be, and in some cases have been, used to derive specific military space technological objectives from broader national and environmental analysis. These range from conventional committee efforts to highly structured—but still intuitive—gaming procedures, such as Boeing's "dynamic contextual analysis," described in Chap. 8. A more systematic approach will be described later.

Environmental analysis in the nonmilitary sphere is generally much less systematic and less integrally involved in the policy-planning process. The essential inputs are (1) broad technological trends, (2) perceived social needs (or wants) and (3) the characteristics of existing socio-political-economic structures and institutions. If there are noteworthy trends in the social area (e.g., changes in sexual mores) these too should be taken into account where relevant. Again, the basic technique is to construct "future worlds" or scenarios whose various implications can be traced and evaluated.

Although one might pick out particular geographical areas and political-military situations, as in the military case, these variables do not seem to be the most salient ones (except for certain specialized industries such as air transport and tourism). Instead, it has typically been found useful to focus on particular societal functions or activities, such as raw-materials processing, services, consumption, communication, transportation, and monetary exchange.

An interesting and provocative "future environment" focused on demands and requirements for information has been sketched by T. J. Rubin, based on a rather generalized picture of society as a system of increasingly interconnected institutions, industries, households, etc. [3]. Each new link in the system, whether it involves goods or services, also involves an exchange of information ranging through shopping lists, price lists, timetables, votes, weather reports, blueprints, publications, stock market data, the U.S. Census, and so on. Rubin particularly notes the growth of specialized "information brokerages" to supply the characteristic information requirements generated by each link, as well as the sharply downward trend of information processing and transmission costs. From such considerations he strongly concurs in the existence of a strong movement toward the development of massive "information utilities" [4], along the lines currently being explored by Western Union (and others), based on the use of large central computers with a great many simultaneous remote input-output stations in different locations. Rubin argues that integrated centralized data-processing and information systems may lead to restructuring institutional forms and boundaries. He ventures five specific forecasts:

1. A renewed impetus toward vertical integration (from manufacturing to distribution) in industry, on the Sears, Roebuck model.

2. An increased degree of government control and supervision of high-information institutions and activities.

3. An increased degree of influence for "organizing" activities—based on information linkages—such as government, banking, and distribution.

4. The importance of transportation will decline in the long run *vis-à-vis* communication. Business travel, in particular, will (eventually) decline.

5. The existing trend towards consolidation and integration in business will continue, with fewer and fewer firms dominating the economy.

Implications of this "future environment" for the computer industry, the communications industry, the U.S. Post Office, the air transportation industry, and others are derivable from the broad picture. Specific goals or functional objectives are not, of course, logical consequences of the scenario itself, but such goals can be determined far more intelligently in consideration of the scenario—and others—than otherwise. Thus corporate policy for a company in the computer or communications field might be to "ride" the trend by exploiting the business opportunities foreseen in the scenario. An organization which perceived itself to be threatened by such a development might decide to oppose it, for instance, by petitioning the FCC to exert very tight controls. An airline would certainly wish to consider the scenario very seriously before embarking on long-range capital investments in supersonic planes—although it might still conclude that the investments were justified.

The actual choice of goals and policies for the organization or institution necessarily remains a question of intuitive judgment, because criteria for choice cannot be derived indefinitely from prior assumptions. Only *after* the fundamental aims and objectives are decided or agreed upon is there any possibility of developing meaningful or consistent decision rules. At some stage the logical regression from conclusions to premises must reach an end: by definition and common usage that end (or starting point) is at the basic policy level.

STRATEGIC PLANNING METHODS

At the strategic level, the problem of planning has two phases: (1) elaboration of a set of alternative approaches or paths to achieving a functional objective and (2) selection from among them.   For the first phase the primary tool is *mission taxonomy*, a version of morphological analysis described in Chap. 5.   Briefly, an effort is made to delineate all—or as many as possible—of the ways of achieving the goal.   As Linstone points out, it is important to develop strategic concepts, or missions (in military parlance), *outside* the menu of conventional responses [5].   In practice, it may be advisable to arbitrarily exclude the standard approaches, at least for purposes of the exercise.

Although the language and terminology here have been developed, in the first instance, for military planning purposes, it is obviously applicable with only minor changes to many business or financial situations.   Indeed one hears not infrequently of "price wars," "invasions" of new sales territory, and similar expressions.

The process of selection from among various alternatives involves questions other than technical performance, but it is possible *in principle* to derive a formula or a formalism by which the selection can be made automatically once certain parameters are determined.   In practice, as will be seen, it is usually necessary to introduce a considerable amount of human judgment in fixing these parameters for two reasons.

1. Objectives are seldom fully and completely defined at the policy level; there are almost invariably significant residual ambiguities, uncertainties of interpretation, and semantic confusions.

2. A precise decision rule implies quantifiability and reproducibility.   However, for the present—and the indefinite future—variables having to do with human responses to external phenomena[4] are hopelessly unquantifiable.   There is no exogenous standard in terms of which they can be either defined or measured.

In principle, it would be desirable to extremize some quantitative measure related to *cost-effectiveness,* e.g., to maximize the probability of success achievable for a given cost, or minimize the cost

---

[4] Variables such as comfort or prestige or beauty, for instance.

of designing a strategy with a specified—say 90 percent—chance of success. Actually, the above examples are illustrative of a number of possible *cardinal* measures of utility which might be used, depending on the situation. However at the strategic planning level—unlike the tactical level—such cardinal measures are almost invariably too difficult to define or compute satisfactorily. Moreover, they are not really necessary, since all that is really needed to choose a strategy is an *ordinal* measure (namely, a simple rank ordering of the possibilities).

Indeed, an alternative formulation of the distinction between strategic and tactical planning might be that in the former one aims at achieving ordinal measures of utility, whereas in the latter a cardinal measure like benefit-cost is required. This problem will be discussed in the next section.

Although an ordinal measure of utility is what is desired and needed for strategic planning, the typical practice among scientific planners—especially in the military—is to develop a pseudocardinal measure, based on a hybrid combination of quantifiable parameters such as dollar cost and ordinal figures of merit such as "military worth," measured on an arbitrary scale of, say, 0 to 10. This approach was pioneered by Churchman and Ackoff in 1954 [6] and first applied to military problems by Howard Wells in a 1958 thesis [7].

$$\text{System desirability} = \frac{\text{military worth} \times \text{feasibility}}{\text{cost}}$$

Evidently, one can invent almost any number of qualitative relationships of the general form

$$\text{Desirability (of a project)} = \frac{\text{utility (if successful)} \times \text{chance of success}}{\text{effort required}}$$

The method can be generalized to cover the situation where a set of options must be compared on the basis of a number of *different* characteristics or in terms of simultaneously satisfying a variety of different and possibly incommensurate objectives. Further complexity may be introduced if the relative importance of the various objectives is not immutable, but itself varies, depending on the external circumstances (i.e., the scenario). There is no single logical resolution of this problem: one approach often followed is to combine (by summation) the various criteria into a single composite one by introduction arbitrary relevance numbers or "weight factors" (on a scale of 0 to 1), which must also be determined independently. Another approach, which has become identified with *linear programming,* is to set minimum standards of compliance with the various objectives as absolute constraints (i.e., inequalities) on the variables. A criterion for choice might then be: that system (or project or strategy) whose composite desirability index is the highest *and* which simultaneously satisfies all requisite minimum conditions or constraints. Examples will be given later.

Obviously the numerical values attached to some of the generalized factors going into a "desirability index" or a priority number have no intrinsic significance: they are only meaningful in relation to one another. Thus, if the potential utility or worth and the relative chances of success of 10 different options are rated on an arbitrary scale, and divided by an appropriate measure of the effort required to achieve each one, one obtains 10 different numerical values of "desirability." These 10 values will not have any particular quantitative significance, unless two are quite close together, but their *order* will be significant. It is possible to assert that the resultant ordering is the most rational one possible, *based on the given information.*

The final ordering of systems or strategies does depend on the cardinality of the factors.[5] Hence it is important to estimate the latter as carefully and objectively as possible—despite their vague definitions and ambiguous associations. Originally it was thought adequate to determine such numbers by polling experts. Recently more and more sophisticated methods have been used to achieve meaningful group consensus on such issues. Feedback techniques similar to Delphi and even man-machine "games" are now employed to this end.

Evidently the numerical rating scheme described above can be applied to almost any problem involving the selection of one out

---

[5] It will be observed that the factor ratings can be changed, without altering their order, in such a way as to alter the order of the resultant.

of a large number of possibilities. Questions like "Which college should I choose?" "Which house should I buy?" "Which candidate should I vote for?" "Which job should I take?" can be analyzed in terms of the above framework. However the approach is particularly well suited for guiding decisions involving a large number of technological alternatives: weapons systems, new product developments, research projects, etc. It can be explained and discussed best by describing a particular model for facilitating such choices which was developed in 1963–1966 by the Military and Space Science Department of Honeywell Corporation [8–10], based on the ideas propounded in 1958 by Wells. The model, described below, is known as PATTERN.[6] Its purpose is to assist in planning at the strategic level, but in accomplishing this it spans the entire spectrum of levels down to specific hardware developments, starting with an environmental analysis or scenario and a set of explicit overall national objectives, and proceeding in any of a variety of directions, as defined by the interests of the planners (e.g., military/space, medical technology, urban transportation). The output of a model "run" is a set of relevance numbers (or relative priorities) for listed projects, systems, or technologies.

The model is based on a detailed hierarchical set of relationships, called a *relevance tree,* based on morphological considerations. (This notion was first introduced in Chap. 5.) At the highest level of the tree (the policy level) is a set of national goals. In a 1966 test of the model in a military/space application, for instance, three political military objectives for the United States were postulated:

National survival . . . . (weight: 6)
Credible posture[7] . . . . (weight: 3)
Favorable world opinion . . . . (weight: 1)

By means of a consensus poll, given independently to two sepa-

[6] An acronym for *P*lanning *A*ssistance *T*hrough *T*echnical *E*valuation of *R*elevance *N*umbers.
[7] Not fully explained by the model designers, but probably interpretable as (1) maintaining forces-in-being fully consistent with stated national commitments and (2) maintaining a reputation for living up to commitments in order to reassure allies and minimize the chances of miscalculation by an opponent.

rate groups, it was determined that these three objectives (treated as independent variables) should be weighted in the ratio 6:3:1. Despite the close agreement between the averages for the two groups, this procedure is clearly a weak point, as already noted.[8]

Subsequently, the procedure followed was to identify three basic types of national *activity*,

Active hostilities (military)
Noncombat (intelligence, arms control, etc.)
Exploration

Nine coefficients were assigned, again by consensus polls, representing the relative importance of upgrading the national capability to carry out each of the three major activities from the standpoint of achieving each of the three objectives. By averaging over the three objectives (weighted as above) they came up with the numbers in parentheses as weighting factors for the activities, in Table 9.2.

The next step was to break down the 3 basic activities into 13 smaller categories such as global war, local war, counterinsurgency (COIN), exploration-space, etc. Then 64 *missions* were identified covering the entire spectrum of activities; these were subdivided in turn into 204 tasks. To carry out these missions the analysts came up with 697 primary *systems concepts* and 2,368 secondary systems involving several thousand different *functional subsystems*. Since there were sometimes several different hardware approaches, one group of 425 functional subsystems involved about 850 competing subsystem *configurations*, excluding those already

---

[8] The fact that the analysis restricted the poll to "experts" is a tacit admission that the outcome would probably depend on the degree of sophistication (in national security affairs) of the participants—not to mention their sociopolitical orientation. It is interesting that upper-middle echelon industry and government executives tend to attach a relatively low importance to "world opinion." However, this may well be a comment on the parochialism of their outlook, rather than an accurate representation of the national or top leadership consensus on these matters. It may be significant that a series of United States presidents before Lyndon Johnson, faced with actual decisions in crisis situations, have vetoed the proposals of military leaders on political grounds in which world opinion was a major, if not dominating, ingredient. Choosing relative weights for national objectives is an extraordinarily difficult task at best; for anyone who is not both sensitive and "tuned in" on all the relevant wavelengths it is probably impossible.

TABLE 9.2

National activity	Policy-level criteria			
	Ensuring national survival (wt. 0.6)	Demonstrating credible posture (wt. 0.3)	Creating favorable world opinion (wt. 0.1)	Final weight
Military..............	0.6	0.6	0.4	0.58
Noncombat...........	0.3	0.1	0.1	0.22
Exploration..........	0.1	0.3	0.5	0.20

in production, design, or far advanced in development, and a large number (2,000) of specific technological prerequisites yet to be achieved (*deficiencies*) were pointed out.

The complete relevance tree can be used to identify all objectives at any level of interest which would benefit from achievement of (or, conversely, would suffer from failure to achieve) any particular task at a lower level. By symmetrical arguments one can easily identify all tasks, subsystems, or components which contribute to a given mission or objective.

As a first approximation, of course, all connecting lines are drawn with equal emphasis, and contributions either "exist" or do not exist. However a natural extension of the method would be to attach indications of relative importance (i.e., "shades of gray") to the connecting lines. The problem of quantification thus broached is similar in kind to the assignment of weight factors to various objectives at the top level.

Thus, appropriate criteria for choice were defined for each level. Those used for the top level have already been illustrated. At successively lower levels criteria such as threat, force structure, capability, prestige, cost-effectiveness, operational advantages, requirements met, feasibility, risk, and scientific implications were applied. At each level a matrix is set up to match various options (activities, missions, tasks, systems programs, etc.) against various criteria, for the particular future environment or scenario under consideration.

The matrix representation of the relationship between successive levels is a convenient computational tool which permits one to compute composite relevance numbers spanning more than one level by simple matrix multiplication. Disregarding the question of the variable importance of different connecting lines—shades of gray—the matrix method can be described quite simply. Suppose L_1 and L_2 are two levels. (L_2 being the higher one) having m and n elements respectively. Let the elements of L_1 be listed vertically and those of L_2 horizontally, and at the intersection of each row and column put a "0" if there is no line between the corresponding elements of L_1 and L_2 and a "1" if there is such a line. The result is an $m \times n$ matrix of 0s and 1s. The extension to the case where connecting lines are not all equally important is immediate and obvious: the major links are still assigned the number 1, but others are assigned appropriate smaller numbers in proportion to their relative importance.

The complete tree is vastly more detailed than the schematic hierarchy shown in Table 9.1. Thus a slightly abbreviated version of the military/space tree described above is shown as Fig. 9.3 [11]. A relevance tree devised by the National Aeronautics and Space Administration for the Apollo Program is shown in Fig. 9.4 [12]. Approaches very similar to PATTERN have been or are being applied to strategic decision making in a number of industries and organizations, notably by Swager (Battelle Institute) [13] and Cetron (Headquarters Navy Materiel Command) [14,15]. A recent survey by Cetron, Martino, and Roepcke lists 30 quantitative models applicable to the selection of research and development projects [16].

COST-EFFECTIVENESS

A measure often used in the military planning—both at strategic and tactical levels—and equally applicable wherever a functional goal can be defined is *cost-effectiveness*. Although it is quite common to use this term as though it were a parameter to be used for systems or programs rating ("high" versus "low"), it actually refers to a technique for explicitly exhibiting the tradeoffs between sunk costs and results achieved. In general the technical perfor-

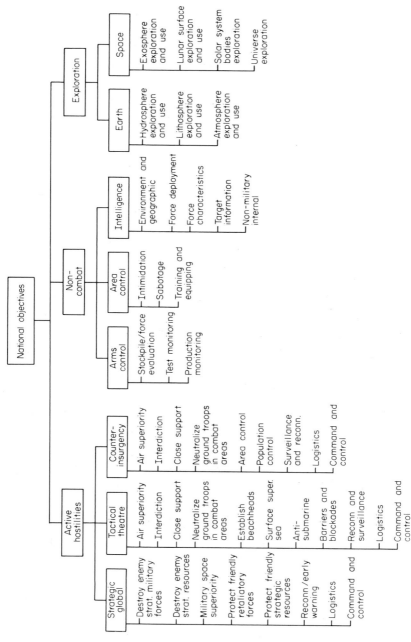

figure 9.3 *Honeywell's military space relevance tree.*

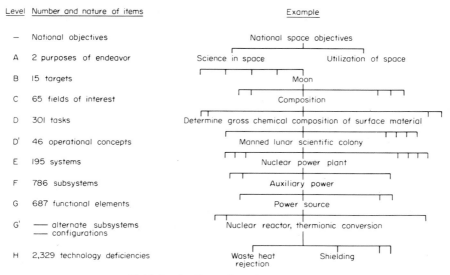

Level	Number and nature of items	Example

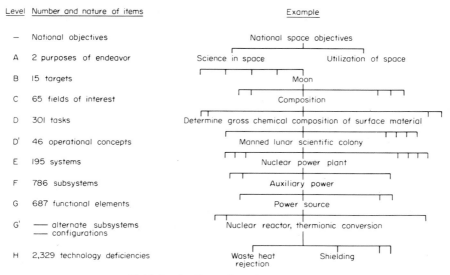

figure 9.4 *NASA's Apollo payload-evaluation relevance tree.*

mance—or effectiveness—of a system or program is a function of how much money is being spent on it: with respect to any given level lower performance can be achieved at less cost, and better performance will cost more. These variations are not linear, however. As one approaches the limits imposed by the state of the art at any given time, small increments in performance tend to be inordinately expensive.[9] On the other hand, at low performance levels it may be possible to achieve a considerable upgrading by a relatively modest additional expenditure. Evidently cost-effectiveness is not a simple parameter but a functional relationship, as shown in Fig. 9.5.

Even in a go/no-go situation[10] where the objective is fixed and it is inappropriate to describe performance effectiveness in terms

[9] For instance, the costs of using chemically fueled rockets to put a man on the moon are excessively high because there is virtually no design leeway, due to the severe constraints on payload. If NASA had been willing to contemplate using nuclear explosives for propulsion—as was proposed and proven to be technically feasible by the Air Force's Project Orion as early as 1958—the entire Apollo mission could have been accomplished for about $5 billion [17].

[10] e.g., to put a man on the moon by 1970.

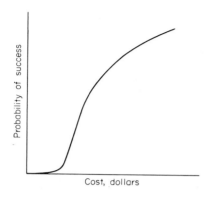

figure 9.5 *Cost effectiveness for go/no-go situations.*

of a range of possible achievements, cost-effectiveness is a perfectly valid concept as applied to a development program. In such a case the appropriate variable is probability of achieving the specified goal, as illustrated in Fig. 9.6.

Where a choice must be made from among a large number of alternative projects, cost-effectiveness can be a useful strategic planning tool. Optimization criteria may be any of the following (among other possibilities):

1. To maximize "military value" within a fixed budget
2. To maximize the probability of achieving a designated capability within a fixed budget
3. To minimize the cost of achieving a designated capability
4. To maximize technological cross-support and spinoff while achieving a minimum designated goal within a fixed budget
5. To do any of the above subject to additional constraints (such

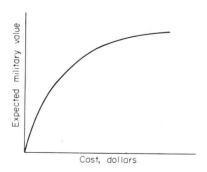

figure 9.6 *Cost effectiveness for declining marginal utility situations.*

as maintaining in-house capabilities, minimizing level of support for "sacred cow" projects or institutions)

Quantitative methods of optimizing for any of the above are touched on briefly in the next chapter under the heading *operations analysis*.

Cost-effectiveness considerations are used in strategic planning as a means of helping to decide what the broad program—or the project "mix"—should be. It can be used at the tactical level in another more sophisticated way to minimize the cost or development time and/or maximize the performance of a specified system based on specified subsystems and components. This will be discussed later.

THE PLANNING-PROGRAMMING-BUDGETING SYSTEM (PPBS)

Starting in 1961 the U.S. Department of Defense under Robert S. McNamara introduced a modus operandi which became known as the PPB system [18]. Since 1965 the system has been spreading to other government agencies under the influence of the Bureau of the Budget[11] [19]. The system is not a planning technique as such, but it is worth discussing briefly because it has stimulated

[11] Thirty-one agencies were using the system by 1967.

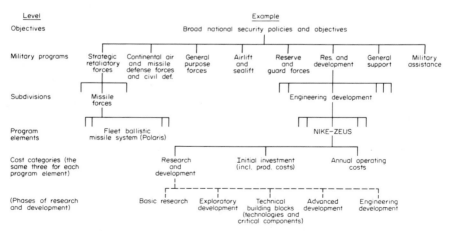

figure 9.7 *PPBS decision tree—military.*

a good deal of interest in, and development of, quantitative methods. It has also been instrumental in focusing continuous attention on the evolution of the relationship between operating programs and the goals which they are supposed to further.

The PPB system as originally set forth consists of three stages:

1. Mission-oriented military requirements determination
2. Formulation and review of programs
3. Preparation of annual budget estimates

The process starts with broad national policies as determined by the White House. This provides basic guidelines for the sliding 5-year Joint Strategic Objectives Plan (JSOP) prepared annually by the Joint Chiefs of Staff and planners in the respective services to relate force requirements to military missions. The Office of the Secretary of Defense has the responsibility of translating force requirements into resource requirements. In civilian agencies a comparable procedure is followed. During this stage systems analysis and long-range technological forecasting play a very prominent role in the military case and are becoming increasingly important in the civilian agencies.

The second phase of PPBS is actually the implementation of the objectives (i.e., the plan), which are prepared annually on a (sliding) 5-year basis, together with cost assessments based on systems-analysis studies done in the previous phase. Both strategic and tactical planning are encompassed here. The final result is an annual "Five-Year Force Structure and Financial Plan" issued under the signature of the Secretary of Defense. The third phase—the budget—is taken from the first year of each successive 5-year financial plan.

The PPB system is explicitly built around a 4-level decision-tree concept which is closely related to the relevance trees discussed earlier. Two examples, military and civilian, are illustrated in Figs. 9.7 and 9.8 [20].

DEMAND-ORIENTED PLANNING

Up to this point we have emphasized planning approaches in a situation where specific goals are formulated, and where technologi-

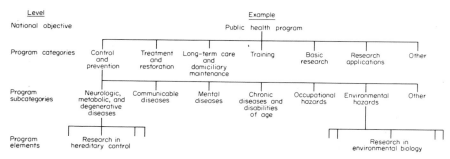

figure 9.8 *PPBS decision tree—civilian.*

cal objectives have been "derived" from the overall goals by some decision process, hopefully an explicit one. For industry in the nondefense sectors, however, it is often unhelpful, and may be misleading, to use this framework. Of course, a corporation may set specific goals for internal growth, but in its economic relationships with the outside world it is normally involved in a competitive market where the goals or requirements of individual customers are important only insofar as they contribute to an overall *demand*.[12]

In a competitive situation, manufacturers are frequently faced with questions such as whether to enter an existing market in which other companies are already established (and, if so, how) or whether to attempt to create a market for a totally new product or service (and, if so, how). In either case, at some stage the product, or product line, must be designed. There will of course be an interplay and feedback among corporate planners, design engineers, market analysts, and production men. If the process is successful, the resulting design concept is one which matches a recognized need (i.e., a demand) and an existing technological capability. When all is accomplished, of course, one has also achieved a contingent technological forecast, in precisely the same sense that an operating plan is such a forecast. To the extent that demand reflects a real social need, such a forecast has a normative component (i.e., latest demand is what *ought* to ex-

---

[12] This does not apply, of course, where a single customer dominates (i.e., *monopolizes*) the market and is in a position to control the specifications of the product. The Defense Department is normally in this position *vis-à-vis* its suppliers, but so are many large marketers such as Sears, Roebuck & Co., A&P, Bordens.

ist).[13]   Where the product-design procedure reveals a demand which cannot be met adequately in terms of existing technology, of course, it produces powerful incentives for undertaking appropriate research and development.

Undoubtedly many corporations use fairly explicit techniques for product design, although they tend not be described in the open literature. However, Battelle Memorial Institute has pioneered in this area, and the following brief discussion is based largely on the work of the members of the Technical Planning Group at Battelle.

The first step is to formulate a fairly precise characterization of the demands of the market being considered in terms of a *needs profile* [21]. It is not very illuminating to describe a market in terms of catchwords or conventional product names. This point is best illustrated with the help of an actual example: Table 9.3 is a needs profile for a hypothetical truck-cab air-conditioning unit [22].

TABLE 9.3

Characteristic	Value index
Low maintenance cost	.35
Low power requirements	.25
Low initial cost	.16
Reliability (low down time)	.13
Air-distribution comfort	.05
Low bulk	.04
Other	.02
Total	1.00

The weight factors are not based on measurements: they are subjective estimates of relative importance, based on a survey of trucking company purchasing agents, normalized to add up to unity. They play much the same role as Wells's measures of military "worth" described earlier, and were obtained by using the Churchman-Ackoff procedure [23].

---

[13] There is sometimes a confusion here, especially with regard to consumer goods, between designing new products and designing new "images" of old products. The latter is the advertising business. It is pertinent to planning and forecasting to the extent that image building can alter people's responses to objective reality and thereby influence it. However this topic will not be discussed here.

The "needs profile" helps to focus the design effort on the right things. In the above case, for instance, low maintenance cost and reliability count for many times more than the drivers' comfort to the purchasing agents of the trucking companies—who are the actual customers. Again, as alternative design approaches are proposed, they can be compared in terms of how closely they match the theoretical demands of the market [24].

The relevance of technological forecasting becomes evident when one notes, perhaps, that the refrigeration technology which would rank highest in terms of low maintenance cost, reliability, *and* low power requirements is one which utilizes waste heat from the engine. But this implies an entirely different class of thermodynamic devices from conventional vapor-compression cooling systems. Cheaney points out that one now asks, not *whether* novel systems concepts will evolve to meet this need, but *under what circumstances* such an evolution might occur. At this point a systematic morphological scrutiny of all possible thermodynamic cooling systems might be warranted, followed by an analysis of the necessary and sufficient technological conditions to achieve feasibility. The result might be a decision to move ahead with a particular approach, or—as likely—a decision to hold up pending some intermediate development, such as a drastic increase in the heat transfer capability of materials. The observed rate of increase of the salient performance parameters, as a function of time, provides a convenient benchmark for estimating the time when a novel product-development approach might successfully be undertaken. Of course, if the organization is large enough and the market potential is sufficiently great, an in-house research and development program might be undertaken to upgrade the state of the art.

REFERENCES

1. Dennis Gabor, *Inventing the Future,* Martin Secker & Warburg, Ltd., London, 1963.
2. Erich Jantsch, "Technological Forecasting in Corporate Planning," delivered at *2nd Annual Technology and Management Conference,* Washington, D.C., Mar. 18–22, 1968.
3. Theodore J. Rubin, "Technology, Policy and Forecasting," *Proceedings of the 1st Annual Technology and Management Conference,* Prentice-Hall, Inc., Englewood Cliffs, N.J., 1968.

4. Martin Greenberger, "The Computers of Tomorrow," *Atlantic Monthly,* July, 1964.
5. Harold Linstone, "On Mirages," *Proceedings of the 1st Annual Technology and Management Conference,* Prentice-Hall, Inc., Englewood Cliffs, N.J., 1968.
6. C. W. Churchman and R. L. Ackoff, "An Approximate Measure of Value," *Journal of the Operations Research Society of America,* vol. 2, no. 2, May, 1954.
7. See Howard Wells, "Weapon System Planners Guide," *Long Range Forecasting and Planning,* Symposium at U.S.A.F. Academy, Aug. 16–17, 1966.
8. Aaron L. Jestice, "Project PATTERN," Joint National Meeting, Operations Research Society of America, Minneapolis, Oct. 7–9, 1964.
9. J. V. Sigford and R. H. Parvin, "Project PATTERN: A Methodology for Determining Relevance in Complex Decision-Making," *IEEE Transactions on Engineering Management,* EM-12, March, 1965.
10. Erich Jantsch, *Technological Forecating in Perspective,* OECD, Paris, 1967.
11. Jestice, *op. cit.*
12. *Missiles and Rockets,* Apr. 7, 1966.
13. William L. Swager, "Technological Forecasting and Practical Business Plans for Petroleum Management," Battelle Memorial Institute, November, 1965, cited by Jantsch, *op. cit.*
14. Marvin J. Cetron, "PROFILE—Programmed Functional Indices for Laboratory Evaluation," 16th Military Operations Research Symposium, October, 1965.
15. Marvin J. Cetron, "QUEST Status Report," *IEEE Transactions on Engineering Management,* EM-14, March, 1967.
16. M. J. Cetron, J. Martino, and L. Roepcke, "The Selection of R&D Program Content—Survey of Quantitative Methods," *IEEE Transactions on Engineering Management,* EM-14, March, 1967.
17. Theodore B. Taylor, "Propulsion of Space Vehicles," in *Perspectives in Modern Physics,* Interscience Publishers, New York, 1966. Cost estimate from an Air Force proposal to the Office of the Secretary of Defense (T. B. Taylor, personal communication).
18. Charles J. Hitch, *Decision Making for Defense,* University of California Press, Berkeley, Calif., 1965.
19. Henry S. Rowen, "Improving Decision-making in Government," Summer seminar on Systems Analysis and Program Evaluation, U.S. Bureau of the Budget, 1965; cited by Jantsch, *op. cit.*
20. Jantsch, *op. cit.*
21. J. A. Hoess, "A Discipline for Both Obtaining and Evaluating Alternative Product Concepts," ASME paper M-66-MD-87, January, 1966.
22. E. S. Cheaney, "Technical Forecasting by Simulation of Design," in *Long Range Forecasting and Planning,* symposium sponsored by the Air Force, Aug. 16–17, 1966.
23. Churchman and Ackoff, *op. cit.*
24. E. S. Cheaney, "Technical Forecasting as a Basis for Planning," ASME paper 66-MD-62, April, 1966.

10 PLANNING AT THE TACTICAL OR OPERATIONAL LEVEL

In the last chapter it was pointed out that a tactical plan is concerned with implementing a specified strategy or reaching a well-defined technological objective. The *sine qua non* of a good plan at any level is that it optimizes (or extremizes) some index of performance—or *objective function*. At the strategic level an ordinal measure (i.e., a rank ordering) of possible alternatives is sufficient: what is optimized is the process of choice. On the other hand, a tactical plan seeks to optimize an objectively quantifiable performance parameter. For instance a tactical plan for a national economy[1] generally focuses on achieving a set of production targets in each industry, corresponding to specified growth rates. For the economy as a whole, performance is normally measured

---

[1] Such as the 5-year plans or 7-year plans typically promulgated in communist countries.

184

in terms of gross national product (GNP), which is defined as the sum of all goods and services produced—and paid for—including services supplied by government. This is equivalent to the total monetary income received by individuals from all sources.[2] This measure has several well-known deficiencies, including the fact that unpaid services (e.g., by housewives) and goods produced and consumed by households or on family farms or exchanged by barter are omitted, as are "free goods" such as water and the assimilative capacity of the environment. Also, GNP is not by any means a reliable measure of welfare or economic efficiency: it has been pointed out that if a producer generates a great deal of some harmful waste and the government cleans it up, the GNP is artificially increased thereby, but total welfare is not. Unfortunately economists have not yet devised a more satisfactory measure of performance.

In industry there are, again, several possible performance indices of which the most widely used is "return on capital investment." This is not, however, a uniquely defined concept (as will be seen), and there are a number of different methods of measuring return. In a slightly broader context this comes under the heading of *benefit-cost* analysis. Costs are, of course, fairly well defined and straightforward in most instances, but not so benefits. In the case of private industry it is only necessary to consider benefits to the corporation, although in some cases a very broad view may be taken as to what constitutes a benefit. (For instance a well-publicized act in the public service may be of incidental advertising value to the perpetrator.) Regarding public investments, on the other hand, it is necessary to identify all the beneficiaries, as well as all those who may be injured. In too many instances, unfortunately, benefits are computed on the basis of considering only a fraction of the effects on part of the population, and ignoring the rest.[3] For example, it is more or less standard practice for state highway departments and the Bureau of Public Roads to compute the benefits from highways only for road users, totally disregarding many hidden or secondary costs affecting other segments of the

[2] Adjusted for inflationary factors.

[3] In particular, there is a widespread tendency to overlook "externalities" or third-party effects [1].

population.[4]   Methodologically this narrow approach is indefensible for a government agency spending public money, although politically it seems to be advisable.

Even if the attempt is explicitly made to take into account all benefits and all costs to all affected parties, there is another fundamental methodological difficulty in comparing incommensurate values.   The Planning-Programming-Budgeting System (PPBS) described in Chap. 9 has forced many unwilling federal executives to attempt to justify their existing programs in quantitative terms. Although the impact on the bureaucracy has probably been salutary in the sense of stimulating a good many agonizing reappraisals, it has also resulted in some very unsatisfactory measures of utility and/or disutility.   For instance, a benefit-cost analysis carried out in 1966 by the Office of the Assistant Secretary for Program Coordination of the Department of Health, Education, and Welfare (HEW) [2], attempted to project a dollar value on the motor vehicle accident control program for the years 1968–1972.   This encompasses two logical steps of extreme difficulty:[5]

1. A quantitative assessment of the number of fatalities and injuries which would occur if there were no program, and

2. An evaluation of the worth in dollars of the resultant lives saved

The latter evaluation was made by invoking some assumptions which could not possibly be justified, except by the argument that no better methodology exists.   In short, the dollar value of a human life was simply equated to lost lifetime earnings plus direct (out-of-pocket) costs of medical treatment, except that the present value of future earnings was discounted by 3 percent per year (in the same study the present value of future expenditures was discounted at 6 percent; the discrepancy was not explained).   The annual value of a housewife was equated to the mean earnings

---

[4] In many cases there are alternative transportation systems which might have been built instead: a proper accounting of differential effects would have to include relative safety, quantities of pollutants produced, noise, psychic costs of congestion, patterns of land development, etc.

[5] Obviously there are comparable difficulties in assessing the "value" of medical, antipollution, educational, antipoverty, highway beautification and many other programs.

of a domestic servant, which amounted to $2,767 in 1964. One paradoxical result of this method of calculation is that the "value" of a man or woman increases with age to about age twenty-five where it peaks and subsequently begins to decline. Also, contrary to the "women and children first" tradition of the West (but fully in accord with Oriental values), a young woman is worth only about 60 percent as much as a man.

Objections to this approach to estimating benefits are numerous and compelling, but it is much easier to identify its faults than to devise a significantly better measure.

One of the crucial issues in benefit-cost analysis—touched on above—is the method of calculating the *present value* of a future income stream. The question, in another form, is: What should one assume for the effective forward discount rate r? In words, the discount rate is essentially the expected rate of appreciation of alternative investments currently available. The present value of a future benefit is the amount of money which, invested at r percent per annum, would ultimately produce a yield equal to the projected benefit.[6]

Assuming the appropriate rate r can be established in some satisfactory way, the present value of a future income (or cash-flow)

---

[6] In the case of very long-term projects, such as reservoirs and navigational facilities, whose benefits must be computed over a time horizon extending 50 to 100 years into the future, it is easy to see that the calculated benefit is extremely sensitive to the assumed discount rate: the lower the rate, the greater the benefit. For this reason the choice of the appropriate discount rate to use in benefit-cost calculations by agencies such as the Army Engineers and the Bureau of Reclamation has become something of a political issue. Although theoretically the proper choice should be a matter for economists or scientists to choose on the basis of their best professional judgment, the rate which is used in benefit/cost calculations by all Federal agencies is currently based on the average "coupon" rate paid on all outstanding U.S. Government bonds with > 15 years' remaining life. As of July 1968, this figure stood at 3⅛%. A new proposed rule would substitute the "effective" interest rate on the same bonds, which was 4⅝% as of July 1968. However, the lowest rate obtainable by long-term *borrowers* has never been less than about 4-¼ percent and is currently around 7 percent. Analysis by Irving Fox and Orris Herfindahl of Resources for the Future, Inc. has shown that 64 percent of federal water-resources projects approved in 1962 would not appear justified if a 6 percent discount rate had been used, and 80 percent would have been unjustified based on an 8 percent discount rate.

stream $C(t)$ is simply given by the integral

$$P(r) = \int_0^\infty C(t)\, e^{-rt}\, dt$$

where $C(t)$ is the cash flow (or income) per unit time at future time t.[7]   In general, of course, $C(t)$ is *negative* for the first few years after the initial decision to invest in a new project—whether capital construction or a new product—with positive contributions corresponding to profits coming later on.   A project merely breaks even, in benefit-cost terms, if $P(r) = 0$.   Because of the rapidly decreasing exponential factor e^{-rt}, a project with a large but long-delayed layoff may have a present worth much less than one whose return is smaller but earlier.

The notion of extrapolating a constant discount factor into the future is subject to very serious question, since it is equivalent to assuming finite (albeit decreasing) contributions to the present worth of a project from very remote periods.   The approximation is probably not greatly in error in cases where the discount factor itself is fairly large, say greater than 10 percent per annum.   In such a case, assuming a constant annual return for purposes of argument, the total contribution to present worth from all the years after 1983 (if the calculation were done in 1968) would be only 22.3 percent of the total.   On the other hand if one assumes a 4 percent discount factor the years after 1983 contribute 54.8 percent of the total benefits.   If the assumed returns (or "profits") do not even begin for a number of years, as is the case for most water-resources projects for instance, then nearly the whole of their present value must be attributable to the relatively remote future.

It is clearly essential, then, to incorporate an environmental analysis and a technological forecast into any meaningful benefit-cost calculation which depends explicitly, or implicitly, on discounted future returns—the more so the lower the assumed discount rate.   Unfortunately this procedure is almost never followed in practice in regard to public investments: in the rare instances of benefit-cost analysis where technology is given any consideration

---

[7] This formula is mathematically quite convenient, since $P(r)$ is the Laplace transform of $C(t)$. For many specific functions $C(t)$ one can simply look up the corresponding function $P(r)$ in a table [4].

at all, it is usually assumed to be unchanging. Thus the peak of canal building in the mid-nineteenth century occurred at a time when traffic was being lost to railroads; railroad and trolley building in the United States in turn reached its peak in the first two decades of the twentieth century in total disregard of the rising threat of competition by the motor vehicle. It may well be that the vast new interstate highway system will suffer a similar loss of patronage as huge "airbuses" and car-rail "piggyback" services begin to cut into the need for cross-country driving.

The field of water-resources development, especially, has been characterized by very long time horizons and low discount rates, with little or no consideration of the impact of technological change on the assumed benefits. It is clear, for example, that if the anticipated benefits of a multipurpose dam assume a requirement for irrigation water, the benefit-cost calculation should—among other things—include an explicit evaluation of the potential impact of new technologies for water conservation. This may involve some excursions into rather remote areas (at first glance), such as the recent development by American Oil Co. of a technique for "waterproofing" cropland[8] from beneath by injecting a layer of asphalt at a depth of several feet below the surface. This prevents subsurface runoff and, where soils are optimum, increases the water supply available to plants. Obviously it might correspondingly reduce the demand for irrigation water in some cases.

One topic which has consistently produced an immense amount of controversy is the question of benefit-cost analysing of the research and development activity itself. More will be said about this in the next chapter. However it seems worth remarking at this point that there is apparently a definite trend in industry and government to erase the traditional distinction between research and development and other types of investment. To put it another way, research and development is coming to be seen as a normal and necessary part of the operations of every corporation (or government agency) which has any stake in future change—and, of course, very few organizations do not have such a stake. But if

[8] The method is not universally applicable; but then there are often other methods for other situations.

research and development is a normal cost of doing business, should it not be expected to make regular and programmable contributions to the organization's objectives? Both the Defense Department and private industry have increasingly begun to answer a firm "yes" to this question. In many cases an attempt has been made to go further and calculate the present value of the future benefits of research and engineering development within the discounted cash-flow framework.

It is difficult to quarrel with the notion that research and engineering projects should be ranked in terms of some sort of "present value" criterion. The difficulty—immediately recognized by many critics of the procedure—is that many, if not most, of the really important long-range research and development projects of the past would have had a low priority on this basis, as compared with many short-term "quick-fix" and service-engineering tasks—especially if the discount factor is taken to be fairly high (as it normally should be). Although most organizations have sensibly not been seduced into throwing out their long-range research programs, some have found themselves rather at a loss for a good rationale for defending them.

Here, again, the difficulty arises from uncritically assuming a constant discount rate r for all purposes. It must be remembered that the real significance of such a discount is that there always exist other opportunities for investment which will produce a return of r percent. But this implicitly assumes a constant (or increasing) flow of new technology to replace that which will be made obsolete in the interim. We have noted already that it is not satisfactory to assume a constant discount factor in the very long run for calculating the present worth of a fixed investment such as a canal in an environment of changing technology, because the underlying technology itself will eventually be outmoded. It is still less rational to attempt to assess the present value of scientific research by any procedure which is predicated upon continuing technological progress and certainly implicitly presumes the continuance of research. To use discounted cash-flow techniques in this application is logically inconsistent. It could even lead to disaster, in that a corporation which consistently optimized on this basis might find itself dominating an obsolescent market, having neglected the most important long-range consideration of all: to survive (i.e., stay in business!).

In both the cases discussed above—investment in a fixed-capital asset and investment in research—the present-value formalism can be used (in principle) by making modifications in it.   In the first instance the problem is to find a way of reflecting adequately the progressive obsolescence of the fixed asset *vis-à-vis* investable liquid capital (money) as time goes on.   Mathematically this implies a discount rate which *increases* with time, thereby reducing the present value of benefits realized in the remote future.   By contrast, the problem of evaluating research on new technology is to find a way of reflecting the progressive obsolescence of the current technology, which implies a *decrease* in the effective discount rate as time goes on.   Actually, it is probably safer not to use a discounted cash-flow methodology at all, at least for present-worth calculations of either long-lived fixed-capital assets or long-range research and development projects.   In the latter case, particularly, it is essential that *enough* research and development be done to assure a continuing flow of new technology or new products as needed to meet the organization's objectives.   The implementation of this criterion will be discussed in the next chapter.

OPERATIONS ANALYSIS[9] AND SYSTEMS ANALYSIS [5,6]

The previous section had several peripheral points of relevance to technological forecasting—especially in regard to the calculation of future benefits—but its central thrust was to prepare for a discussion of optimization, which will be undertaken hereafter.   To the extent that operations analysis is concerned with "optimizing" a programmed sequence of actions leading to a new system or functional capability, it may be thought of as a form of self-fulfilling technological forecast.   Optimization is actually much too strong a word, however, since it implies a universal measure of utility and a specified future environment.   Neither, in general, can safely be assumed.   Indeed operations analysis and systems analysis are

[9] Also called Operations Research or OR.

both heavily concerned with *suboptimization* of a set of distinguishable (but not orthogonal) variables simultaneously with respect to several different criteria, in an uncertain environment.[10]

The subject matter of operations analysis and systems analysis is vast and includes a large number of topics not particularly relevant to technological forecasting. It also encompasses much that has been dealt with briefly already in this book (e.g., decision trees). It is clear, of course, that planned actions undertaken to achieve an objective depend on decisions which are likely to be influenced above all by considerations of cost. Methods of minimizing project costs through efficient allocation of resources, for example, cannot be considered totally irrelevant to the study of technological change, since comparative projected costs are likely to determine which projects are chosen in the first place. However at this point we feel justified in restricting the discussion to methods of estimating and minimizing the time required to achieve a planned goal (and the associated problem of scheduling), since—as noted earlier—such analysis obviously is a form of contingent forecasting.

In this connection, undoubtedly the logical starting point is "network analysis," which is hardly more than the art of displaying the causal relationships among a set of abstract events in graphic form. If one is analyzing the requirements for a project whose end point is well defined, the nodal points ("events") of the network are specific *activities*. The basis for such an analysis is to

1. Identify all pertinent activities
2. Identify their causal relationships
3. Draw the appropriate interconnections as directed lines on a chart

The lengths of the lines have no particular significance. The logically possible causal relationships between two events A and B are as follows: (1) A is a prerequisite of B, (2) B is a prerequisite of A, or (3) none. In the latter case A and B are independent and can be carried out simultaneously or in any order which happens to be convenient.

---

[10] Hereafter when the word optimization is used, it will be understood in this very restricted sense.

TABLE 10.1 Activity List for Shock-tube Project

Activities in this column —	Must be preceded by these activities —	And must be followed by these activities —	The total time in days for each activity is —
P Preliminary study...........	...	A1, B1, C1, D1	6
A Foundation			
A1 Design slab and buttress..	P	A2, A3, A5	7
A2 Grading and forms......	A1	A7	6
A3 Acquire reinforcing steel..	A1	A4	4
A4 Fabricate reinforcing steel	A3	A7	6
A5 Design tube supports.....	A1	A6	1
A6 Fabricate tube supports..	A5	B8	2
A7 Place reinforcing steel....	A2, A4	A8	1
A8 Pour concrete and set....	A7	B8	4
B Tube			
B1 Select tube.............	P	B2	2
B2 Acquire tube...........	B1	B6	8
B3 Design flanges..........	C1	B4	3
B4 Acquire flange material..	B3	B5	4
B5 Machine flanges.........	B4	B6	6
B6 Weld flanges to tube.....	B2, B5	B7	5
B7 Post-weld machining.....	B6	B8	10
B8 Assembly on foundation..	A6, A8, B7	D3	1
C Diaphragms			
C1 Study and selection......	P	B3, C2, C4	7
C2 Acquisition.............	C1	C3	12
C3 Preparation.............	C2	E1	3
C4 Striker design...........	C1	C5	3
C5 Striker fabrication.......	C4	E3	5
D Controls and instruments			
D1 Selection...............	P	D2	3
D2 Acquisition.............	D1	D3	10
D3 Installation.............	B8, D2	D4	2
D4 Checkout and calibration	D3	E1	2
E Facility checkout			
E1 Static pressure test.......	C3, D4	E2	2
E2 Determine diaphragm burst pressure.........	E1	E3	10
E3 Calibration with striker..	C5, E2	...	5

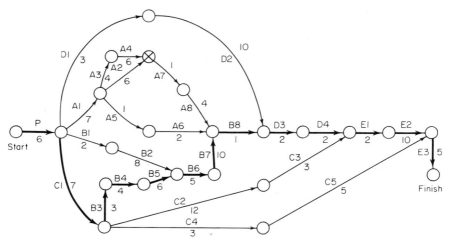

figure 10.1 *Activity network for shock-tube project.   Time estimates for each activity (in days) allow duration of each possible path, or sequence of tasks through the network, to be determined. Longest path is that shown in heavier line.   This critical path sets the minimum length of the project.*

As an illustration, Table 10.1 is a detailed list of the activities involved in designing and constructing a small experimental shock-tube facility [7].   Logical predecessors and successors are identified in the first two columns.   When these relationships are expressed graphically the result is as shown in Fig. 10.1.   Time estimates for each activity, derived from consultations with specialists, are listed in the last column on the right.   In a more complex project involving large numbers of different activities, or many repetitions of the same activities, the simple estimates above may be replaced by "expected" times based on statistical distributions of actual times based on many trials.   The latter scheme is used in PERT (*Program Evaluation and Review Technique*), a computerized elaboration of the network approach which was developed originally to monitor the Polaris submarine project [3].

Once the activity times are specified—either deterministically or probabilistically—the minimum total time for the project can be found by comparing the elapsed time of all paths through the network, and identifying the longest.   This is the so-called *critical path*, which is drawn with a darker line than the others in Fig.

10.1 [9]. In this diagram it is 63 days. On all other paths there is some leeway for adjusting the starting time for particular activities to allocate scarce manpower resources most efficiently, or to satisfy other constraints. The definition of this leeway—called *slack time*—for starting a particular activity is the difference between the earliest and latest possible starting times. The first is the time of the longest path preceding and leading to the activity; the second is the time of the critical path minus the time of the longest path leading from the activity to the end point. For instance if we consider the node marked by an X (off the critical path) marking the beginning of activity A7, it can be verified quickly by inspection that the earliest starting time is 23 days after the beginning of the project and the latest is 36 days after the beginning (or 27 before the end.) Hence there are 13 days of leeway for starting this particular activity. The power of the method is revealed by the ease with which such questions can be answered.

Evidently the minimum length of the project may be longer than the critical path length if additional constraints are imposed. For instance, the diagram in Fig. 10.1 assumes there will be no delays due to shortages of labor, work space, computer time, machine shop time, etc. More elaborate models can take such constraints into account. It is also possible to treat the case where several projects must be handled concurrently within a set of overall resource constraints[11] [10]. In this situation it is not possible, in general, to simultaneously optimize each project separately; but in principle one can minimize any desired composite function of the individual project times (e.g., the average or weighted average, the longest time, the average times, the variance).

As described in the foregoing paragraphs, the so-called *critical path method* (CPM) is primarily applicable to projects involving relatively pedestrian engineering developments, at most. If the project goal requires substantial departures from the existing state of the art, of course, the problem of estimating times for the intermediate stages must be regarded as a nontrivial exercise in tech-

---

[11] Indeed the "executive" system for a time-sharing computer system must cope with precisely this problem.

nological forecasting, whence the critical path estimation—indeed
the entire network analysis—would also be a contingent technological forecast of what could be done if maximum effort were put
forth.

Where the technological goal is far beyond present capabilities—
as, for instance, controlled thermonuclear power or three-dimensional (holographic) color TV—the network ceases to be determi-
nate. Several, or perhaps many, nodes will be preceded and/or
followed by a set of alternative paths. The former implies the
existence of several possible ways of solving a problem; the latter
implies the existence of several possible choices depending on the
outcome of an activity. The identification of all possible alternatives in either case is a task for morphological analysis, which was
discussed in Chap. 5.[12]

It was pointed out in the last chapter that the cost-effectiveness
framework is applicable also at the tactical level as a sophisticated
optimization tool. This may be seen most clearly in the case of
an elaborate system with a number of operational subsystems each
involving some degree of departure from the current state of the
art. The Apollo moon project is an overworked but still appropriate example. Typically certain performance goals are essentially unchangeable, due to exogenous circumstances. In the case
of Apollo one such fixed goal is minimum payload capability, fixed
by the nature of the mission, the physical characteristics of human
beings, and the necessity of providing for their sustenance during
the trip. Another immutable Apollo requirement is safety: since
humans were to be the passengers there was almost no room for
compromise on this score.

In Apollo, however, several subsystems contribute to each requirement, and it is possible (in principle) to make some tradeoffs
between various subsystems. For instance, payload in orbit could
be increased in a variety of ways, such as by using more energetic
fuels in the first and second booster stages, or by adding an additional stage, by extreme microminiaturization of on-board equip-

---

[12] If attention is focused specifically on the problem of selection among
alternatives, the network is often described as a *decision tree* or a *relevance
tree*. These are important conceptual devices for strategic planning (Chap.
9).

ment, or even by sending two capsules into a moon orbit, one manned and one unmanned.[13] (Extra payload would also contribute to the overall margin for safety during the mission.) On the other hand, several of the alternatives which might increase overall payload would add greatly to the cost as well as the complexity of the undertaking. *Ceteris paribus,* greater complexity tends to be correlated with reduced reliability, with an adverse impact on safety.

In sum, then, there were conceivably a number of ways of achieving the requisite payload and safety. The task of systems analysis in this case might be to minimize the overall project cost by choosing appropriate combinations of targets based on the estimated cost-effectiveness relationships of the various subsystems— subject to the above constraints. Or, in the case of Apollo, it might have been judged more important for political reasons to minimize the overall time required to achieve the objective.[14] These alternative criteria would, in general, imply quite different development strategies. Thus a cost-minimizing strategy would usually involve making a choice, in advance, of the most promising line of development and following it on a "business-as-usual" basis. Time minimizing, on the other hand, requires pursuing parallel lines of development simultaneously on a "crash" basis.

Even when the overall system configuration has been settled and frozen, however, the planner is often faced with a choice between pushing the state of the art to the extreme limit in a single subsystem versus demanding a smaller departure from present capabilities in several different subsystems. The nonlinear nature of cost-effectiveness relationships (see Figs. 9.6 and 9.7) immediately implies that it is both cheaper and will take less time to achieve a 10 percent improvement in two subsystems of roughly equal cost than

---

[13] The unmanned capsule could be a reserve source of fuel and/or additional electronic navigational assistance, etc.

[14] The mathematical techniques used might be *linear programming, quadratic programming,* or *dynamic programming.* It does not seem appropriate to digress from the theme of this book into a discussion of these mathematical topics, even though the latter two are not particularly well known except among academic OR specialists. In any case, the real-world applicability of currently well-developed mathematical methods is rather limited and it appears that heuristic approaches are often of equal or greater value.

it is to achieve a 20 percent improvement in a single one. Of course the balance between improving three subsystems by 10 percent *vis-à-vis* improving a single one by 20 percent is not obvious and can only be settled by reference to actual cost-effectiveness curves. The complex series-parallel interrelationships of networks such as Fig. 10.1—and all the more so for a project as complex as Apollo or Polaris—requires a highly sophisticated extension of the above type of comparison, generally with the assistance of electronic computers. However network analysis, combined with the cost-effectiveness approach, offers a methodological framework for optimizing time to completion, cost, reliability, operational effectiveness, or any other parameter or combination of parameters of interest to the system designer. Clearly, also, an optimization method can be used as a forecasting tool (with an important normative or teleological component) based on the assumption that planners will themselves attempt to perform such an optimization.[15] With the rapid spread of scientific management such an assumption would seem to be increasingly justified, although its range of validity is probably still quite narrow.

FORECASTING ERRORS AND
THE PLANNING LOOP

Many students of management have discovered (or rediscovered) the information "feedback" cycle between planning and performance. This observation is essentially a recognition of the adaptive character of the process. One schematic representation of the relationships is shown in Fig. 10.2 [11].

In recent years especially there has been a tremendous amount of development work on quantitative procedures for research and development project selection [12,13]. Despite the large number of models in existence, very few are extensively utilized in practice. The reason, as Meadows points out, is that the rationale for the selections is not subject to external validation of any kind [14]. In most cases the only test of validity is the internal consistency and

[15] For instance, the U.S. Defense Department may forecast Soviet military accomplishments, or a corporation may forecast what its competitors will do.

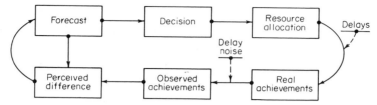

figure 10.2   *Decision feedback loop.*

"reasonableness" of the various quantification and comparison procedures. Not surprisingly this criterion is usually only fully satisfactory to the model designers. Even internal consistency is sometimes surprisingly hard to achieve: various different rating methods often lead to inconsistent results, as illustrated in Table 10.2 [15]. Some of the discrepancies are quite striking.

TABLE 10.2   Attempts to Rank Order Priorities of Military Research and Development Projects by Several Methods

Project	Priority class assigned by Combat Development Command (CDC)	Quantitative value ratings assigned by Army Materiel Command		
		Method 1	Method 2	Method 3
1	1	10.00	9.65	1.90
2	1	9.00	8.20	10.00
3	1	9.75	6.40	9.00
4	1	8.95	4.60	5.30
5	1	8.65	2.80	3.35
6	1	8.35	2.60	5.90
7	2	7.25	8.00	7.75
8	2	6.95	7.50	7.35
9	2	6.95	7.50	7.35
10	2	6.95	7.50	7.35
11	2	4.45	6.40	9.75
12	2	5.85	10.00	8.60
13	2	5.85	9.85	1.00
14	2	3.95	2.05	2.65
15	3	3.10	6.60	8.20
16	3	2.55	4.80	9.35
17	3	1.70	4.05	4.70
18	3	1.00	1.00	4.10

The usefulness of a project-selection technique is nearly impossible to test objectively unless a substantial number of the alternative projects are actually undertaken, and their relative "value" as seen after completion compared with their ratings as determined beforehand. At the minimum such an experimental test would require 4 or 5 years; it would, doubtless, be illuminating, but—so far as the author is aware—no such investigations have been reported to date. However several "before and after" studies have been carried out, notably by Dennis Meadows, of the accuracy of forecasts of (1) cost and (2) probability of success of some 144 research and development projects in industry, spread out among five different (profitable) companies of different sizes and in different businesses [16].

Results of the study showed that cost estimates tended on the whole to be too low for all projects, but especially so for those which were either technically or commercially unsuccessful. This is perhaps not very surprising, since it reflects a fairly common situation where success constantly seems "just around the corner," but—despite continued or even enhanced efforts—it remains elusive, often because of quite subtle difficulties. As a result the research organization continues to pour good money after bad, at least until the hidden joker is identified. There was a very low—almost negligible—correlation between actual and prior estimates of project costs, probability of technical success, and probability of commercial success. There was, however, a definite correlation between cost overruns and the source of the project. Thus projects suggested by customers were notably more profitable and more predictable—presumably because they involved less departure from the existing state of the art—than projects initiated within the research and development laboratories. Projects initiated by the sales departments fell in between on both counts. Results are summarized in Table 10.3.

A study of 26 military research and development projects by A. W. Marshall and W. H. Meckling explicitly divided them into three categories based on the degree of advance beyond existing technology [17]. Here, again, the ratio of average actual to estimated costs varied from 1.4 for minor incremental charges through 1.7 for intermediate cases to 3.4 for major advances.

TABLE 10.3   Percent of Projects Yielding Various Results

Source of idea	No sales increase	Small sales increase	Modest sales increase	Large sales increase	Average ratio of actual to estimated cost
Laboratory.....	66	17	17	0	2.20
Marketing......	58	14	14	14	2.02
Customer......	33	33	13	21	1.27

A study by Irwin Rubin of government-sponsored research and development projects involving expenditures in excess of one million dollars each seems to provide some grounds for optimism in regard to the future of quantitative management methods [18]. Of the 37 projects examined, 20 were controlled and monitored by PERT and 17 used other (presumably less systematic) decision procedures for allocating resources and scheduling activities. The frequency of cost overruns dropped from 71 percent among the non-PERT-controlled projects to 40 percent in projects where PERT was used.

REFERENCES

1. R. U. Ayres, and A. V. Kneese, "Production, Consumption and Externalities," *American Economic Review* (in press), 1969.
2. U.S. Department of Health, Education, and Welfare, "Application of Benefit-cost Analysis to Motor Vehicle Accidents," Office of Assistant Secretary for Program Coordination, August, 1966.
3. Irving K. Fox and Orris C. Herfindahl, "Attainment of Efficiency in Satisfying Demands for Water Resources," *American Economic Review,* May, 1964 (Resources for the Future, Inc., reprint no. 46).
4. See, for instance, C. J. Tranter, *Integral Transforms in Mathematical Physics,* John Wiley & Sons, Inc., New York, 1951.
5. See R. L. Ackoff, E. L. Arnoff, and C. W. Churchman, *Introduction to Operations Research,* John Wiley & Sons, Inc., New York, 1957.
6. R. L. Ackoff, *Scientific Method: Optimizing Applied Research Decisions,* John Wiley & Sons, Inc., New York, 1962.
7. Gale Nevill and David Falconer, "Critical Path Diagramming," *Modern Science and Technology,* D. Van Nostrand Company, Inc., Princeton, N.J., 1965 (reprint from *International Science and Technology*).

8. See D. D. Roman, "The PERT System: An Appraisal of Program Evaluation Review Technique," *Journal of the Academy of Management,* vol. 5, April, 1962.
9. Nevill and Falconer, *op. cit.*
10. For instance, see Burton V. Dean and L. E. Hauser, "Advanced Materiel Systems Planning," *IEEE Transactions on Engineering Management* EM-14, no. 1, March, 1967.
11. Dennis L. Meadows, "Characteristics and Implications of Forecasting Errors in the Selection of R&D Projects," presented at the 2nd Annual Technology and Management Conference, Washington, D.C., March, 1968.
12. N. R. Baker and W. H. Pound, "R&D Project Selection: Where We Stand," *IEEE Transactions on Engineering Management* EM-11, no. 4, December, 1964.
13. M. J. Cetron, J. Martino, and L. Roepcke, "The Selection of R&D Program Content: Survey of Quantitative Methods," *IEEE Transactions on Engineering Management* EM-14, March, 1967.
14. Meadows, *op. cit.*
15. Burton V. Dean and Lawrence E. Hauser, "Advanced Materiel Systems Planning," *IEEE Transactions on Engineering Management* EM-14, March, 1967.
16. Meadows, *op. cit.*
17. A. W. Marshall and W. H. Meckling, "Predictability of Costs, Time and Success of Development," in NBER (ed.), *The Rate of Direction of Inventive Activity: Economic and Social Factors,* Princeton University Press, Princeton, N.J., 1962.
18. Irwin Rubin, "Factors in the Performance of R&D Projects," in *Proceedings of the Twentieth National Conference on the Administration of Research,* Denver Research Institute, University of Denver, 1967.

11 PLANNING
FUTURE RESEARCH

Research and development managers daily face, in a fairly sharp form, the apparent contradiction between *need* or *mission* orientation and *opportunity* orientation. As has been noted earlier, especially in Chap. 3, this duality permeates the entire subject matter of this book. But for the planner himself—one might suppose— there should be no controversy: his job is to bring order out of chaos, to guide, to define, plot, and measure the path of progress toward fixed goals.

But what (or whose) goals must the planner seek to implement? The research worker may be interested in science for its own sake, or for the sake of the prestige and personal advancement it brings. The engineer may be interested in making something that "works." But at the management level, where money decisions are made, almost nobody is concerned with science or technology per se.

Managers, as well as taxpayers and ordinary citizens, are naturally concerned with end results, as measured in terms of national prestige, security (i.e., defense), improved health, better living standards, employment opportunities, or simply profits. Research and development contributes to these objectives, to be sure, but in a way which is not correlated in any simple fashion with the scientific importance of what has been learned. Indeed, the work of some of the most brilliant investigators has no appreciable direct utility to anyone outside the scientific fraternity, whereas the work of a journeyman technician may turn out to have the highest national importance.

It does not follow, of course, that "pure" science should not be supported, since important indirect benefits accrue to society from major centers of creative basic research. Half a dozen Nobel prize winners—or the equivalent—can draw together a galaxy of lesser (but not much lesser) lights, who in turn attract research grants, graduate students, science-based companies, government contracts, and major investments.[1] However, the direct support of pure research projects with no payoff in sight is always somewhat uncertain. In times of cutbacks and economizing these funds are (or appear to be) especially vulnerable. Perhaps because of this, many scientists seek to justify their research on another, more dubious, basis: serendipity.

It is often alleged by academic scientists or philosophers of science (without extensive documentation, alas) that (1) major scientific breakthroughs, such as the *maser,*[2] are typically not made in industry or government laboratories and (2) that such breakthroughs are essentially unpredictable anyway [1–3]. The clear implication of this line of reasoning is that there is little to be gained by attempting to guide research in toward predetermined objectives.

Since many of the "pure" scientists achieve disproportionate fame and influence compared to their "applied" colleagues, this argument is periodically renewed and usually taken seriously. Indeed, at times it has approached the status of official dogma, with

[1] This pattern is clearly evident around Boston (Route 128) and the San Francisco Bay area.

[2] *M*icrowave *A*mplification by *S*timulated *E*mission of *R*adiation.

the result that some relatively bizarre research projects have been sponsored, on occasion, by government agencies.[3] In due course the pendulum tends to swing too far, however, and a reaction occurs. The agencies of reaction may be congressmen who see a political issue in "wasted" taxpayers' dollars, or—as in the Department of Defense under former secretary McNamara—the reaction may be triggered by an attempt to calculate the benefit-cost ratio of research, based on conventional notions of return on investment.

The question of how benefits from research should be calculated is an exceptionally tricky one, as we have already noted (Chap. 10). For the moment, it is enough to remark that if serendipity is truly a significant factor and if breakthroughs are unpredictable and also inevitable, then benefits of basic or pure research, calculated by the usual discounting technique, tend to be underestimated. Moreover, one can easily verify, as several recent studies have done, that the major technological contributions to those weapons systems which have actually been procured and deployed are almost all the result of applied, mission-oriented research [5,6]. If these studies are based on valid premises,[4] the role of serendipity is apparently slight.

Who is correct, then, Townes and Kistiakowski or the Office of the Assistant Secretary of Defense for Research and Engineering? Granted that Project Hindsight was a response to some real abuses, and that Townes et al. may be reacting more to fears than existing or likely research support policies, it is still perhaps not obvious how to reconcile the two views. The essence of the "serendipity" argument[5] is that the really important new capabilities are often achieved unintentionally, i.e., by researchers who are free to follow a promising line of research wherever it might lead them, even

[3] For instance, an antigravity project was at one time funded by the U.S. Air Force.

[4] There is, however, a contrary viewpoint espoused by some critics to the effect that under McNamara the Department of Defense has consciously rejected all major weapons projects involving new (i.e., not "established") technology—thus automatically ensuring the results of the Hindsight study by robbing them of all significance.

[5] Which, as was noted earlier, is *not* the major justification for continuing to support fundamental or pure research.

in directions quite different from their original aims.   Alexander
Fleming's accidental discovery of the antibiotic properties of penicillin—which remained a laboratory curiosity for 10 years—is a
classic case in point.   Kistiakowski notes that nobody could have
foreseen the relevance of the Curies' discovery of radium to the
treatment of cancer, and that a "need" for such a treatment played
no part in the process [6].   Townes's own invention of the maser
was not directed at any particular objective, except building a
very refined form of frequency oscillator or "atomic clock."   He
was not unduly concerned with advancing the art of radioastronomy, and the dream of extending the maser idea to optical fre-
quencies (the *laser*) probably played no significant part in the
original motivation.   Ergo serendipity does in fact occur and has
been an important factor in at least two monumental discoveries
of recent times.

It does not seem to follow, however, that most such discoveries
could not have been made "on purpose": that is, in a program
of directed research.[6]   On the contrary, if antibiotics had been
sought they could have been found and put to practical use much
sooner than was actually the case.   If radioastronomy had had
a high priority in the 1930s or 1940s there is no reason to suppose
a quantum microwave amplifier could not have been invented to
fill the need (in very much the same way that Bardeen, Brattain,
and Schockley more or less deliberately set out to create solid state
electronics).   If a definite need had been perceived for an intense,
coherent beam of light, the laser might have been invented before
the maser, instead of after.

Indeed, the historical evidence is surely consistent with the notion
that the real reason why antibiotics and the maser appeared "by
accident" is that there was no prior requirement for them.   This
is why it was possible for the developer of the first laser to speak
of it as "a solution seeking a problem [7]."   To be sure, the early
availability of the "solution" in both cases may have accelerated
the day of its widespread application.   It could be argued, in the

---

[6] This does not apply to the discovery of radium, which was totally outside
the then-existing framework (or "paradigm") of physical theory regarding
the structure of matter.

spirit of cost-benefit analysis, that the social utility of inventions or discoveries arising out of undirected pure research[7]—if any—is strictly a function of the difference between (1) the "extra" value of having the particular invention or discovery a few years early and (2) the opportunity costs of the undirected research *vis-à-vis* an equivalent amount of applied research in other fields.

Clearly this comparison can never be made quantitatively or in general. The decision on how to split existing resources between pure and applied research remains a question of judgment and intuition and depends on cases. It seems clear, however, that choices will continue to have to be made among competing pure research projects as well as among competing applied projects. In this more limited sphere technological forecasting and related methods may be of considerable assistance.

MEASURES OF VALUE

Assuming it is conceded that society should underwrite a certain (unspecified) amount of nonapplied, nondirected research, for which (by definition) social utility is an illegitimate criterion of desirability, it seems to follow that if choices are necessary at all they must be made solely on the basis of "scientific value." Unfortunately, this term has remained essentially undefined; at least, no definition has been applied to the practical exigencies of decision making. Instead, the nearly universal practice of those responsible for disbursing research grants is to call on consulting experts, advisory boards, or "visiting committees" to decide on the relative merits of a plethora of competing research projects. The disadvantages of this procedure are many: in the first place, committees are generally too conservative, as noted elsewhere. In the second place, this system probably puts younger scientists too much under the control of their elders for the ultimate benefit of science. Thirdly, the power wielded by such committees is a form of patronage, which tends to be used preferentially (like patronage in other fields) for the benefit of protégés, those with whom one is in agree-

[7] Again, as distinguished from other reasons for supporting it.

ment, and those at one's own institution[8] despite generally very high standards of personal and professional integrity on the part of the committee members. In short, the conventional method of deciding who is to get money and who is not would, in the long run—especially as the recent rapid expansion of support for scientific research slows down—seem likely to put a premium on academic respectability and status quo orientation rather than on creativity and originality.[9]

What is the alternative? There is no substitute for expert knowledge of the subject matter. The use of experts somewhere in the chain of decision making is unavoidable. It is not, however, necessary to delegate the decision itself to outsiders, whose professional interests may well be involved at least tangentially in the outcome. It would be an improvement over the present system to devise some quantifiable measures of "scientific value," perhaps using the experts to provide the quantification. One imperfect and limited, but nevertheless salient measure would be: the amount of new knowledge which would be made *accessible* to scientists by the success of the project. Knowledge becomes accessible, for instance, if some limiting factor or constraint is pushed back.[10] A piece of equipment, a technique, or a discovery which would permit experiments under some hitherto unattainable environmental conditions, or which would make it possible to achieve such conditions much more easily (i.e., more inexpensively), would clearly create such a possibility, as would an experiment or a technique circumventing some previous limitation on the observer or the analyst.

Consider an instance: Research in the fifties on new superconduct-

---

[8] This is certainly one of the most important and least discussed reasons for the centralization of "big science" at a relatively small number of institutions. Of course, a Harvard or M.I.T. professor using his position on such a committee to maintain a steady flow of research grants back to his own department can think of plenty of good reasons for doing so (e.g., the desirability of maintaining—or building up—its present standard of excellence, continuing to attract good students and junior faculty), which may somewhat disguise the inequitable aspects of the arrangement.

[9] This situation has prevailed for many years in French science, due to the rigidity of the academic system and the self-perpetuating dominance of the Ecole Normale Superieure and the Ecole Polytechnique.

[10] The metaphorical phrase "new horizons" clearly expresses this notion.

ing alloys such as niobium-zirconium or niobium-tin has opened up, for the first time, the possibility of producing quite high magnetic fields (>100,000 gauss) in a portable apparatus at a price within the reach of hundreds—even thousands—of research laboratories. In effect, the price of a high-intensity magnetic field has been suddenly reduced by two or three orders of magnitude. Thus, not only the few, expensive, high-priority experiments can be performed, but also many low-certainty, speculative, or exploratory tests will be possible in the near future. The effect of strong magnetic fields on crystal structures, semiconductors, complex chemical reactions, or biological organisms, for example, are practically unknown. It is true that we have no fundamental theories of matter to test at this time, and there is no particular reason to expect anything startling, but this may be merely a measure of our ignorance. It would be surprising if such a potent new research tool failed to lead to some interesting, or even dazzling, new discoveries.

Similarly, a manned orbiting laboratory (MOL), or manned orbiting observatory, would offer unique opportunities to make astronomical measurements without the interference of the earth's atmosphere. Astronomical measurements from the earth's surface are impeded by the opacity of the atmosphere, especially in the short-wavelength end of the electromagnetic spectrum (UV, x-rays, gamma rays). The infrared region is largely blocked by water-vapor absorption. In addition, the atmosphere causes scattering and wavefront distortion, even in the visible portion of the spectrum. Hence a platform above the atmosphere would open up tremendous new areas of knowledge.

Since many of the relatively unexplored areas of science lie at environmental extremes, each order-of-magnitude reduction in the dollar cost of approaching one or another of these extremes is likely to open up fertile fields for speculative investigation. Table 11.1 illustrates just a few of the many possibilities.

Combining several extreme environments at once may offer further possibilities. For example, one of the major long-range goals of physics is to achieve high pressures and extreme temperatures simultaneously in the laboratory—thereby more closely simulating conditions inside stars.

TABLE 11.1

Environmental parameter	Physics	Chemistry	Biology and medicine
High temperature	Plasma dynamics, astrophysics	Refractory metals, free radicals, fluorine chemistry	?
Low temperature	Superconductivity, superfluidity, quantum phenomena	Frozen free radicals, ammonia chemistry, methane chemistry	Cryobiology, suspended animation
High pressure, high gravity	Rheology, seismology, vulcanology, crystallography	Geochemistry	High-gravity undersea environments
Low pressure (high vacuum)	Surface physics	Surface chemistry	Space environments
High frequency, high energy flux	Elementary particles, quantum electrodynamics	Radiation chemistry	Radiation biology, genetics
High E field, high B field	Nonlinear optics, magnetooptics, magnetohydrodynamics	Electrochemistry, magnetochemistry	?
Low gravity	Selenology (geology of the moon)	?	Space environments

Another constraint on the progress of science is the capability to observe and manipulate nature (i.e., detect, resolve, discriminate, amplify, modulate, calculate). Thus, certain sciences, e.g., genetics and virology, will be more tractable if (or when) electron microscopes gain an order of magnitude in resolution. Other sciences, such as meteorology, will profit immensely from new methods of acquiring raw data, or of processing it rapidly. (The computer may be thought of as, among other things, a device for "telescoping" time—doing things in less time is equivalent to extending the amount of time available.)

The increment of increased capability can usually be estimated directly from the trend curves of appropriate figures of merit, as for instance: an experiment to achieve a certain temperature (at a specified cost) or to detect and count a new type of elementary particle, in comparison with existing art. The *utility* of this added capability, in terms of what new knowledge can be achieved from it, is a matter for the wisest of the wise. An increment of new capability in the high-vacuum area may mean very little at the moment, since we have no reason to expect that a vacuum of 10^{-15} torr will reveal anything that was not already known on the basis of 10^{-14} torr. On the other hand, the presence or absence of "gravitons,"[11] for instance, could have far-reaching implications for our whole theory of matter—or it could mean nothing.

Admittedly, the cynosure of all scientific value is not simply increased capability to test, simulate, resolve, discriminate, and so forth. In the last analysis the aim of science is interpretation, integration, and explanation—which take place in the human brain after all the data have been collected. A measure which omits these crucial elements would not reflect the overwhelming importance of the work of the great theoreticians and model-builders such as Newton and Einstein. The accession of new data is a necessary but clearly not a sufficient criterion of scientific value. Somehow one must recognize and take account of the purposes for which the data are collected. In this connection it may be useful to recall Kuhn's paradigm of scientific progress [8]:

1. "Normal science" consists of probing, extension, embroidery, and refinement of established conceptual schemes or models.
2. "Crises" occur when the activity of normal science leads to an accumulation of anomalies which do not fit the accepted concepts. A period of turmoil and confusion ensues during which alternative theories or models "compete" with one another.

Kuhn might object to speaking of the "goals" of research, but he would perhaps admit that there exists—in effect—a sort of tropism which attracts scientists into the problem areas which are perceived to be most *critical* in Kuhn's sense. That is, a criterion of scientific value must include a component related to the relevance of the data to the "burning issues" or "crises" of the discipline. Most data merely elucidate or embroider; a critical experi-

---

[11] Discretely quantized gravitational waves.

ment is one which eliminates (i.e., falsifies) one or more of a set of alternative hypotheses. A truly adequate measure of the scientific value of a project would be based on its potential for reducing the number of possible theoretical models capable of explaining a class of phenomena.

The joker in the deck which inhibits one from attaching quantitative priority ratings to experiments, based on "hypothesis-exclusion potential," is that the number of possible theories is uncountable, not only in the practical sense but also in the strictly mathematical sense. This is true because every theory which must explain a finite number of facts can be embedded in an infinitude of *metatheories* which logically include it.[12] A useful experiment relating to any scientific discipline reduces the class of tenable hypotheses from an infinite number to a smaller but *still infinite* number. The criteria by which scientists exclude all but a (countably) few hypotheses are simplicity (Occam's Razor) and elegance. Within these far more severe limitations there may at any time be one, two, three, or a very small number of obvious competing possibilities.[13] However, the criteria of simplicity and elegance are fuzzy and provide no reliable cookbook recipes for counting significant alternatives.

SYSTEMATIC METHODS FOR PLANNING BASIC RESEARCH

Because of the difficulties enunciated above it is only possible to use operations-research methods (described earlier) with a good deal of caution and care in planning basic scientific research. The relevance tree based on physical objects of research is clearly unsatisfactory, since the criterion of "scientific value" offers no guidance in choosing between objects at any level of aggregation. Similarly a "morphological analysis" of all possible experiments— while not beyond the realm of possibility—falls short of providing

---

[12] Analogously, there are infinitely many continuous differentiable functions which will exactly satisfy any finite number of point conditions.

[13] For example, prior to the successful Watson-Crick model of DNA biochemists and biophysicists were undecided as to whether the structure involved two, three or four strands. There were several attempts to construct a three-strand model. The correct number turned out to be the simplest, namely two [9].

actual decision rules. A Delphi type of approach might be somewhat better than the usual committee of experts, but is really no different in kind.

Is there any alternative? Clearly there is none that has the sanction of widespread use or general acceptance. However T. J. Gordon and M. J. Raffensberger of the McDonnell-Douglas Corporation have attempted to develop a "research model," initially to help guide the National Aeronautics and Space Administration in determining an optimum basic research strategy for the science of astronomy, with specific reference to the definition and analysis of the research programs to be conducted from a manned orbital astronomical observatory [10]. The attempt seems to have been at least partially successful and (so the authors believe) may be applicable in other disciplines.

Gordon and Raffensberger introduced a dual relevance tree with two independent branches: experimental and theoretical. The first consisted of a spectrum of currently feasible observations or measurable quantities, derived as described above from morphologial considerations. The authors identified some 3,000 feasible experiments at the lowest level of the tree. The second branch was divided into two major sets of broad problems: (1) relating to the origin (and future) of the universe "in the large," commonly called *cosmology* or *cosmogeny,* and (2) relating to the laws of change "in the small," i.e., *astrophysics.* Figures 11.1 to 11.4 exhibit parts of the GR theoretical relevance tree [11]. As the illustrations make clear, the theoretical branch consists of a structured set of *questions,* more pointed and more specific as one proceeds from level to successive level.

Evidently, also, the organization in terms of questions is essentially an organization chart for the discipline itself. The intention behind the structure was to exhibit alternative broad theories (e.g., steady-state, evolutionary, "big-bang") and, following each, to list their major consequences and implications—in the form of questions—in such a way that answering the questions casts light on the entire theory. Of course very few (answerable) questions are truly crucial to a model in the sense of being able to demolish the entire structure at a stroke. When an elaborate theory can stand or fall on the basis of a single experiment,[14] of course, the implications of success or failure may send tremors through wider

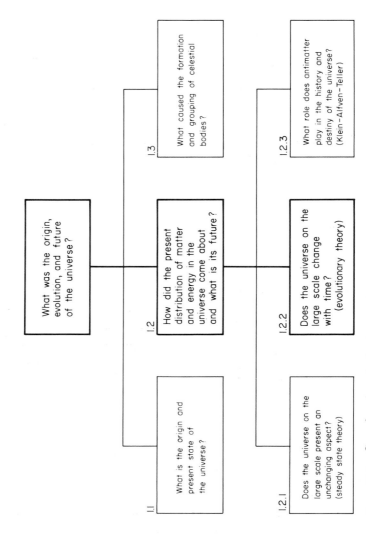

figure 11.1 *Cosmology branch of astronomy tree.*

1.1
What is the origin and present state of the universe?

1.2
How did the present distribution of matter and energy in the universe come about and what is its future?

1.3
What caused the formation and grouping of celestial bodies?

1.2.1
Does the universe on the large scale present an unchanging aspect? (steady state theory)

1.2.2
Does the universe on the large scale change with time? (evolutionary theory)

1.2.3
What role does antimatter play in the history and destiny of the universe? (Klein-Alfven-Teller)

What was the origin, evolution, and future of the universe?

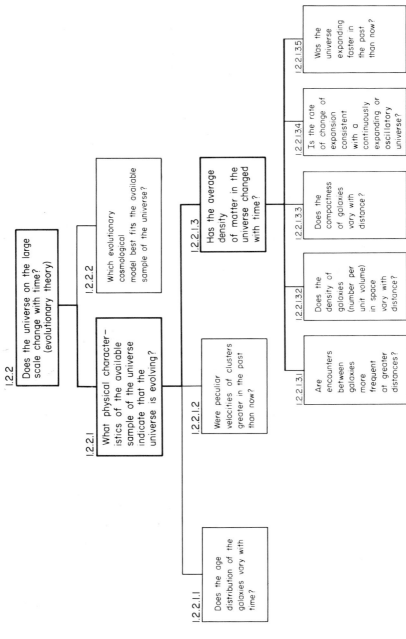

figure 11.2 *Continuation of cosmology tree.*

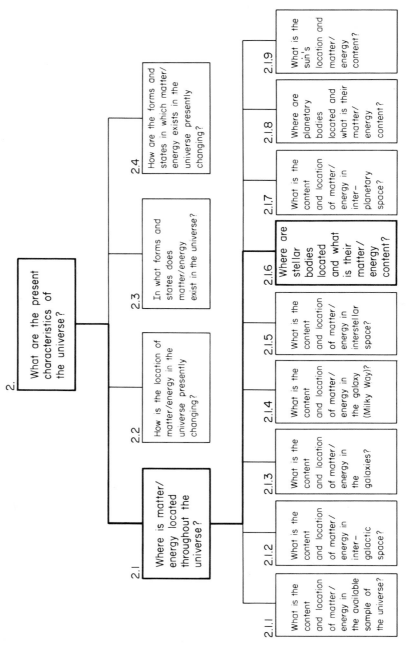

figure 11.3   *Astrophysics branch of astronomy tree.*

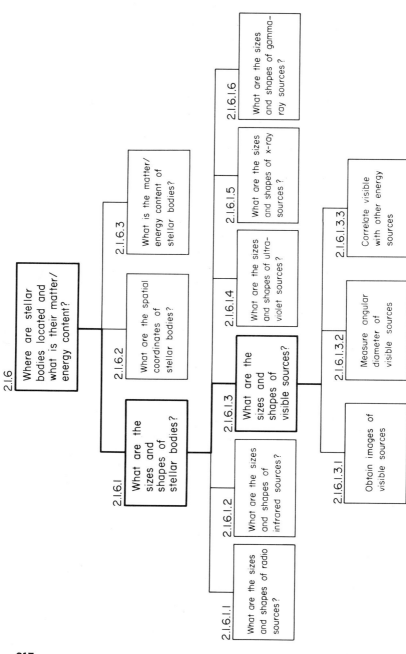

figure 11.4 *Continuation of astrophysics branch.*

spheres than normally encompassed by professional scientists.

The true measure of relevance (or priority) of a basic research project comes when feasible experiments are matched against "crucial issues" which have been derived from the fundamental underlying models or paradigms of the discipline. The more crucial issues probed by a piece of research, the higher its intrinsic priority is likely to be.

There are, naturally, other criteria of practical importance in choosing among research projects. Above all, is there a competent scientist who is enthusiastic about the research? Does it provide cross support for other projects at the particular laboratory or institution in question? A marginal project may well be worth supporting if it is important to a very creative and highly valued researcher. There are also natural constraints to consider: in astronomy certain experiments can only be done at particular times, as when the Earth's orbit most closely approaches that of another planet. Meteorology, bioclimatology, epidemiology, oceanography, vulcanology, archeology, paleontology, cultural anthropology, and several other disciplines can only be pursued effectively in particular circumstances which occur irregularly both in geographic and temporal terms. It is often sensible to take advantage of rare opportunities as they arise, even if other projects of seemingly greater real importance must be temporarily set aside. The Krakatoa study sponsored by the British Royal Society (1888) [12], the work of the Atomic Bomb Casualty Commission in Japan (1945 to ?), and the Alaska Earthquake programs sponsored by the U.S. National Academy (1965–1966) are excellent examples.

RELATING SCIENTIFIC RESEARCH
TO TECHNOLOGICAL OBJECTIVES

While technological forecasting plays an important part in many aspects of the conduct of human affairs, a few of which have been described briefly in the pages of this book, there is probably

---

[14] As the general theory of relativity would have had to be discarded if any of its major predictions—including the bending of light rays in a gravitational field and the precession of the perihelion of Mercury—had proved wrong.

no other area where its role is so central or critical as in the guidance of applied research. Not surprisingly, perhaps, the present efflorescence of technological forecasting as a systematic intellectual adjunct of planning and operations research seems to be closely associated with the tremendous growth of the "research and development industry" following World War II to its present level of about 25 billion dollars a year, or about 3 percent of the gross national product.

In earlier times it was perhaps possible for research administrators to stay abreast of developments in their field and to act, individually, as bridges between the laboratory and the outside world. In principle the administrator was a kind of two-way communications device matching technical needs or requirements from elsewhere in the organization with possibilities emerging from the research effort.

However today—especially in the military—technological needs can only be fulfilled by elaborate complex "systems" involving large expenditures and spanning a number of scientific disciplines. The individual administrator can no longer rely on his own grasp of the enormous subject matter to perform his essential function of expediting, refereeing between alternative approaches, allocating resources, and matching scientific possibilities with technological demands. The addition of staff members to assist the administrator does not entirely solve the problem, for greater breadth of knowledge is brought to bear only at the cost of bureaucratization, reduced flexibility, poor coordination, and increased "noise" in the communication channels.

Increasingly there is a tendency to search for management systems which are less intuitive and more systematic. As much as possible the trend is to break the procedures down into small discrete steps most of which can be performed routinely, so as to conserve and make optimum use of the scarce time of the top policy makers.

Again, the basic analytical tool is the relevance tree, with an appropriate structure. For example, in the military a typical hierarchy would be:

Mission
Mission phase

Technological capability
Basic science
Research project

In an industrial context products and product lines would replace mission phases and missions, respectively. The relevance of a given research project to a particular mission or product line can be analyzed step by step by straightforward methods, variants of which have been described by Dean [13], Smith [14], Cetron [15], and others.

As Fig. 11.5 indicates, between any element on a given level and any element on the next level (either "up" or "down" the tree) there may or may not be a connecting line, representing a positive contribution. Where no line exists there is no contribution. With the aid of such a chart one can identify all research projects bearing on a given mission (or product line) simply by working down the tree from top to bottom. By the same token, one can identify all missions which will be furthered by the success of a particular project by reversing the procedure.

As was mentioned in Chap. 9, the network structure can be represented mathematically by a set of matrices. For instance if l research projects are listed vertically and m basic sciences horizontally, one could assign the number 1 to each row-column intersection corresponding to a positive contribution and the number

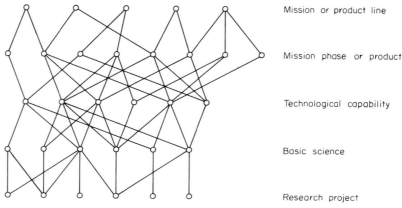

Mission or product line

Mission phase or product

Technological capability

Basic science

Research project

figure 11.5   *Schematic applied research tree.*

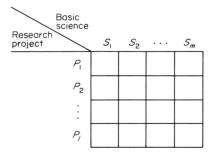

figure 11.6 *Project-science matrix.*

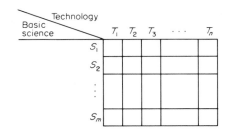

figure 11.7 *Science-technology matrix.*

0 to each intersection corresponding to no contribution, as shown in Fig. 11.6. Clearly, as was indicated in Chap. 9, the numerical entries need not be restricted to 0s and 1s. If contributions of varying degrees of relative importance are allowed, they may be represented by numbers anywhere in the range zero to unity, the latter being reserved for the most important links.

A precisely analogous procedure can be used to relate basic sciences with technological capabilities, technological capabilities with mission phases, and so forth. Thus Fig. 11.7 shows a schematic science versus technology matrix.

The classification of sciences and technologies into exhaustive, nonambiguous, and nonoverlapping categories is a nontrivial problem in itself [16]. It can never be done in any ultimate, time-independent sense. However, some breakdowns are obviously superior to others. The list of basic sciences currently in use in the Defense Department is as shown in Table 11.2.

Classifications of technological capabilities tend to vary much more, depending on the agency or industry. It is clear, however, that one can develop a list as concise or as detailed as desired.

To establish the relevance of specific research projects to technological capabilities the $l \times m$ and $m \times n$ matrices above can be multiplied together to produce a single, composite $l \times n$ matrix relating projects directly to technologies.

$$\begin{bmatrix} \ \}l \text{ projects} \end{bmatrix} \times \begin{bmatrix} \ \}m \text{ sciences} \end{bmatrix} = \begin{bmatrix} \ \}l \text{ projects} \end{bmatrix}$$

$$\underbrace{\qquad}_{\substack{m \\ \text{sciences}}} \qquad \underbrace{\qquad}_{\substack{n \\ \text{technologies}}} \qquad \underbrace{\qquad}_{\substack{n \\ \text{technologies}}}$$

T A B L E 11.2   A Classification of the Basic Sciences

1. Physical sciences
    1.1   General physics
             1.11   Solid state physics
             1.12   Atomic molecular physics
             1.13   Quantum and classical wave physics
             1.14   Acoustics
             1.15   Plasma and ionic physics
             1.16   Theoretical physics
             1.17   Relativity and gravitational physics
             1.18   Quantum fluid physics
             1.19   Instrumentation
    1.2   Nuclear physics
    1.3   Chemistry
    1.4   Mathematical sciences
2. Engineering sciences
    2.1   Electronics
    2.2   Materials
    2.3   Mechanics
    2.4   Energy conversion
3. Environmental sciences
    3.1   Oceanography
    3.2   Terrestrial sciences
    3.3   Atmospheric sciences
    3.4   Astronomy and astrophysics
4. Life sciences
    4.1   Biological and medical sciences
    4.2   Behavior and social sciences

By summing the numbers along the *row* corresponding to a particular project one obtains the number of technologies to which it is relevant, weighted according to the importance of the contribution, if variable degrees of importance are taken into account. By summing the numbers down the *column* corresponding to a particular technology one obtains the number of projects which contribute to it, again weighted by the relative importance of the contributions.

The word "contribution," which has been used repeatedly in the last few pages, refers, of course, to an appropriate measure of utility. But the only appropriate index of worth, at the level of technological capability, is performance itself. In other words the methodology described above is explicitly directed toward predicting the relative impact of any specified collection of research projects—covering an arbitrarily wide span of subject areas in the basic sciences—on any given parametric measure of technological performance.[15]   By an obvious extension of the argument, one also has a potential management tool for choosing program content and allocating budgets in such a way as to match any desired "profile" of technological urgencies.[16]

Apart from such "blue sky" possibilities for manipulating research programs to optimize objectives, the foregoing makes an explicit connection between what is actually being investigated in research laboratories today and the indices of technological performance which will be upgraded thereby.   This does not, so far as it goes, provide any quantitative measure of the *rate* of change of a macrovariable as a function of the relevant level of effort of research.   However a study of rates of change in areas where the level of research support has changed drastically, especially in the downward direction,[17] might help to fill in this gap.

What has been said so far in this section has mainly to do with how much research, divided among which projects, exerts leverage on which technological parameters or figures of merit.   An important issue not yet touched on is *timing*.   Broadly speaking the purposes of applied research are to solve problems or develop products (which may be the same thing).   However because of resource limitations and opportunity costs—which imply that benefits paid for today but received in the future must be appropriately discounted—it is possible to be too early or too late.   A product for which there is no need, or "a solution looking for a problem," is not necessarily worthless, but its future value must be discounted heavily.   Besides, the pioneer always makes more mistakes and has to work harder and invest more than those who follow in

---

[15] In Chap. 6, such measures were called *macrovariables*.

[16] It must never be forgotten, however, that the relevance numbers in the matrices—on which everything else depends—are based on human judgment, not empirical measurement.   The relevance-tree technique is only a means of systematizing and ordering judgment, not of replacing it.

[17] A list of major defense research and development projects which have been cancelled or cut back heavily over the past decade (Orion, nuclear plane, Dyna-Soar, Skybolt) would probably turn up some candidates.

his footsteps.   Hence it can be quite expensive to be first in a new field if one is too early.   It is a fact that early leaders not infrequently fail to gain any permanent advantage from their head start.   An obvious example is Remington-Rand (now a division of Sperry-Rand Corp.), which was first on the market with a commercial electronic computer—the Univac I—but quickly fell far behind IBM in the race for industry leadership.

By the same token, it can obviously be disastrous to be too late. The failure of RCA, GE, Sylvania (now General Telephone and Electronics), Raytheon, and other electron-tube manufacturers to foresee or take advantage of the semiconductor revolution resulted in a permanently lost opportunity.   Industry leadership passed to Texas Instruments Corp., Fairchild, Transitron, Hughes Aircraft, and others.   The development of integrated circuits or "molectronics" was initiated by Texas Instruments and Westinghouse, with Fairchild and Motorola in hot pursuit.[18]

The problem of correct timing has two aspects: (1) a forecast relating to the market or need and (2) a forecast of the availability of the new technology.   These questions have been discussed elsewhere in this book from the methodological standpoint, and there is no need to repeat such a discussion.   However it may be worth recalling from Chap. 6 that technological change in fast-growing fields can be pictured graphically by a series of escalating S-curves (whose envelope may itself be an S-curve).   This escalation process represents the displacement of whole technologies by their successors—precisely what happened when electron tubes lost most of their applications to transistors, which are in turn being challenged by integrated circuits.

Clearly, then, it may be of value to a research administrator to have a tool—even if not very precise—which would help him to recognize when a new technology is approaching the "takeoff" point where it will surpass the old one in terms of performance.

Figure 11.8 graphically illustrates the case of a well-developed "old" technology A on a rapidly rising trend versus a less developed "new" technology B, whose growth curve is still fairly flat.   At

[18] A similar case of permanently lost opportunity was cited by Jantsch: the xerographic process was offered to and refused by most of the major companies in the photocopy business [17].

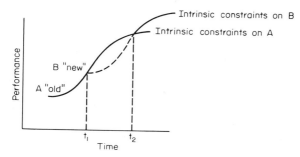

figure 11.8   *Use of S curves to guide timing decisions.*

time t_2 every company will jump on the bandwagon to switch from A to B; the company with foresight enough to steer its research and planning to B at the earlier point t_1—where it first became visible—will gain a substantial advantage over its competitors. This use of growth curves for research planning has been particularly pointed out by T. I. Monahan [18].

REFERENCES

1. Charles H. Townes, "Quantum Electronics, and Surprise in Development of Technology," *Science,* vol. 159, pp. 699–703, February, 1968.
2. George B. Kistiakowski, "On Federal Support of Basic Research," *Basic Research and National Goals,* Report to the Committee on Science and Astronautics by the NAS, March, 1965.
3. Thomas S. Kuhn, *The Structure of the Scientific Revolution,* The University of Chicago Press, Chicago, 1962.
4. Raymond Isenson and Chalmers Sherwin, "Project Hindsight," First Interim Report, October, 1966.
5. Arthur D. Little, Inc., "Management Factors Affecting Research and Exploratory Development," April, 1965 (DDC accession no. AD, 618,321).
6. Kistiakowski, *op. cit.*
7. Theodore H. Maiman, president of Korad Corp. (a subsidiary of Union Carbide) at a press conference. Reported in *The New York Times,* May 6, 1964.
8. Kuhn, *op. cit.*
9. James D. Watson, *The Double Helix,* Atheneum Publishers, New York, 1968.
10. T. J. Gordon and M. J. Raffensberger, "A Strategy for Planning Basic Research," presented at the 2nd Annual Technology and Management Conference, Washington, D.C., March, 1968.
11. *Ibid.*

12. "The Eruption of Krakatoa and Subsequent Phenomena," Report of the Royal Society, London, 1888.

13. Burton V. Dean, "A Research Laboratory Performance Model," *IEEE Transactions on Engineering Management,* EM-14, March, 1967.

14. D. F. Smith, "Long Range R&D Planning," *IEEE Transactions on Engineering Management,* EM-14, March, 1967.

15. M. J. Cetron, "QUEST Status Report," *IEEE Transactions on Engineering Management,* EM-14, March, 1967.

16. J. P. Martino, "A Classification System for Military Functions, Technologies and Sciences," *IEEE Transactions on Engineering Management,* EM-14, March, 1967.

17. E. Jantsch, *Technological Forecasting in Perspective,* OECD, Paris, 1967.

18. T. I. Monahan, "Current Approaches to Forecasting Methodology," in *Long Range Forecasting and Planning,* symposium sponsored by U.S. Air Force, August, 1966.

INDEX

INDEX

SOCIAL SCIENCE LIBRARY

Manor Road Building
Manor Road
Oxford OX1 3UQ
Tel: (2)71093 (enquiries and renewals)
http://www.ssl.ox.ac.uk

WITHDRAWN

This is a NORMAL LOAN item.

We will email you a reminder before this item is due.

Please see http://www.ssl.ox.ac.uk/lending.html
for details on:

- loan policies; these are also displayed on the
  notice boards and in our library guide.

- how to check when your books are due back.

- how to renew your books, including information
  on the maximum number of renewals.
  Items may be renewed if not reserved by
  another reader. Items must be renewed before
  the library closes on the due date.

- level of fines; fines are charged on overdue books.

Please note that this item may be recalled during Term.